A Growers Guide *for* BALANCING SOILS

A Practical Guide to Interpreting Soil Tests

William "Crop Doc" McKibben

Foreword by Don Huber

A Growers Guide for Balancing Soils
A Practical Guide to Interpreting Soil Tests

Copyright © 2021 by William L. McKibben

All rights reserved. No part of this book may be used or reproduced without written permission except in cases of brief quotations embodied in articles and books.

The information in this book is true and complete to the best of our knowledge. All recommendations are made without guarantee on the part of the author and Acres U.S.A. The author and publisher disclaim any liability in connection with the use or misuse of this information.

Acres U.S.A.
P.O. Box 1690
Greeley, Colorado 80632 U.S.A.
(970) 392-4464 • (800) 355-5313
info@acresusa.com • www.acresusa.com

Printed in the United States of America

Publisher's Cataloging-in-Publication

McKibben, William L., 1952–
A growers guide for balancing soils' a practical guide to
 interpreting soil tests / William L. McKibben—1st ed.,
 Greeley, CO, Acres U.S.A., 2021
    vii, 244 pp., 26 cm
  Includes Index
  ISBN 978-1-60173-032-9 (trade)

1. Soils — analysis. 2. Soil testing. 3. Crops — soil fertility.
4. Agriculture — crops & soils. 5. Plant — nutrition.
I. McKibben, William L., 1952– II. Title.
  S593.M35 2021   631.4

Dedication

I dedicate this book to my Dad, a World War II veteran, who died July 25, 2018, and who helped me in my business many a day. Thanks, Dad.

Preface

I AM WRITING THIS BOOK BASED ON MY EXPERIENCES AND knowledge collected over forty years of crop consulting. I am sure there are many things that could be added to this compilation of knowledge, but this book represents my refined approach to balancing soils. I adhere mainly to the Albrecht philosophy of balancing soils using the basic cation saturation ratios; however, my experience has shown me that this philosophy is not applicable in some situations and that using the sufficiency levels of available nutrients is a more realistic approach. Sometimes it is a combination of both methods. It is my goal to help you find the best method for your situation.

I hope to cover in this book changes that I have seen during my tenure as a soil consultant and what farming may look like in the future. My degree from The Ohio State University is in soil chemistry, but there is so much more to growing crops than just manipulating nutrients. Albrecht's approach to balancing soils was to adjust the soil calcium and magnesium ratios in order to provide the best flocculation of clays. This resulted in the best possible soil structure to maximize water infiltration and soil aeration. This also created the best soil environment for proliferation of beneficial soil microorganisms. This balancing approach requires that the soils have a sufficient amount of clay; but what happens when soils are composed of mostly sand or silt that do not carry a negative charge like clay? A different approach must be used.

In today's urban and container farming, the soils are manipulated so much that they are no longer a true soil, but a modified soil mix. These soils fall into a category of their own, and the testing and interpretation of these results need to be adjusted.

No-till farming has become a mainstay in the agricultural community and this has resulted in significant changes in the nutrient profile of our soils. Those issues will need to be addressed in the near future if water quality is to be improved.

Irrigation water quality has a significant impact on soils and nutrient availability. The quality of water will affect the yield and the quality of whatever crop is being produced. Rainwater quality has changed significantly over my lifetime as a consultant, and it too is affecting nutrient availability.

The acceptance of tissue analysis, and now sap analysis, ranges from a total waste of time to the greatest thing since sliced bread. These analyses, when used with the right supportive tools, can be invaluable for heading off nutrient deficiencies. Tissue analysis used to be somewhat questionable, but with the demand for bigger and better yields, most soils will not supply the necessary nutrients at the right time and in the right amounts during the growing season.

I hope to address all these things and many more with a commonsense approach to maximizing crop production and quality. This will not be a book just for the organic farmers or the commercial farmers. I hope to apply my experience and knowledge so that no matter what production method you choose, you will find benefits for your operation.

Acknowledgements

THIS BOOK WOULD NOT BE POSSIBLE WITHOUT THE CLIENTS AND people that I have worked with over the past decades. My clients' farms have been my proving ground and research facilities. As I look back I sometimes wonder why my clients stuck with me, when I compare what I knew then to what I know now. And yet as I write this book, I know that there is so much more to learn. It is my hope that this knowledge can be built upon and improved over time.

Logan Labs has been my lab of choice, and it has truly been a pleasure to work with such fine people. They have afforded me the opportunity to perform in-field research by running countless analysis, including soil, paste, tissue, and water. All the soils data in this book has been prepared by the Logan Labs staff, an effort that is greatly appreciated.

Foreword

AGRICULTURE, AS THE BASIC INFRASTRUCTURE OF SOCIETY, IS responsible for providing the necessities of an abundant, affordable, and nutritious source of essential food, feed, fiber, and other commodities. The vast majority of these commodities are produced on small family farms and in home gardens where the localized ecology is managed to nourish and enhance the plants grown. There is great personal satisfaction from fulfilling a personal responsibility in growing your own necessities and achieving that measure of self-sufficiency and confidence. Balancing mineral nutrition for optimum results is a key part of this production because minerals are constituents of plant tissues, as well as regulators, inhibitors, and activators of physiological processes. As plant constituents, carbon, hydrogen, oxygen, nitrogen, phosphorus, magnesium, calcium, and sulfur are required in large quantities (macronutrients) to produce the carbohydrate, protein, and lipid components of the cells and tissues. Potassium and the micronutrients (cobalt, copper, iron, manganese, nickel, and zinc) primarily function as regulatory components for the physiological processes critical to production of the nutrient-dense, high-yielding, healthy products society relies on for food, feed, and fiber.

Each nutrient functions as part of an integrated system with the plant's genetics and environment. Efficient production requires a balanced nutrition to fulfill the regulatory and composition processes involved. Deficiency of an essential nutrient has domino-like repercussions throughout the system, just as an excess can result in toxicities through interference with the function of other elements. Each plant type has its own specific requirements, and few environments provide an optimum balance of all nutrients for all plants. The availability of many nutrients for plant uptake may be environmentally determined rather than a matter of presence or absence. Thus, balancing nutrition for optimum production requires a knowledge of the specific plant's needs at various growth stages, nutrient availability, and a favorable environment for their function.

In other words, gardening is managing the ecology to optimize the availability of nutrients consistent with the plant's needs. This management may require adding specific nutrients to soils where they are absent, modifying the environment to increase nutrient availability when present but unavailable, or reducing toxic levels when present in excess.

Visual symptoms of nutrient deficiencies often occur long after initial yields and quality are reduced. This is especially true for the "hidden hunger" that occurs with micronutrient deficiencies. Soil testing is a powerful tool to tell us what is present, and plant tissue analysis tells us what the plant can find that is available for uptake. Understanding your environment, and effective use of soil and plant analysis, can provide a solid basis for producing high-yielding, nutrient-dense produce and weed-, disease-, and pest-free food for your family's needs and enjoyment. To assist in this endeavor, William "Crop Doc" McKibben has shared his forty-plus years of research and experience in *A Grower's Guide for Balancing Soils* to make nutrient management a key to unlocking the potential production and joyful reward that comes through a vigorous and well-managed garden. It will be a valuable reference and guide for the home as well as commercial gardener who wants to reap the benefits produced through the miracle of plants.

— *Don M. Huber,*
Professor Emeritus of Plant Pathology, Purdue University

...

Bill McKibben's first book, *The Art of Balancing Soils*, has been an inspiration to countless young and experienced agronomists across numerous industries. As a mentor, he has given so many of us a clear focus on the value of the Albrecht model of base saturation, even as it continues to be under attack by an uninformed community. His simple and easy-to-read style takes a potentially complicated conversation to a place where it can truly be used to create success in an era where misdirection can be costly.

In his follow-up book, *A Grower's Guide for Balancing Soils*, Bill continues our understanding of how a soil really works. It's his ability to bring us back to basics that is so important. He lets us see the significance of how soil chemistry, physics, and biology have to work together in order to produce a quality crop, regardless of what that crop may be, and he makes it very clear that chemistry is our first limiting factor.

In *A Grower's Guide for Balancing Soils*, Bill continues to share with all of us his depth of experience. This book will become so much more than a simple guidebook. *A Grower's Guide for Balancing Soils* is an important read for all students of good soil management.

— *Joel Simmons,*
EarthWorks

...

This is the book that lets you see the forest through the trees. Bill has devoted his lifelong career to studying, educating, and rethinking agriculture and soil science. His common-sense approach and desire to truly help people is refreshing. This book will provide a roadmap to practical and sustainable solutions for novices and experienced growers a like.

It has been my privilege and honor to work, learn, travel, and conspire with the "Crop Doc." Bill has been a great friend to Logan Labs and one of the greatest assets to our clients. This work, along with *The Art of Balancing Soil Nutrients*, are must haves for all growers.

— *Susan Shaner,*
Lab Director, Logan Labs

Table of Contents

Preface vii
Acknowledgements ix
Foreward xi

CHAPTER 1
The Value of Soil Testing 1

CHAPTER 2
Exchange Capacity, pH, and Organic Matter . . . 11

CHAPTER 3
Nitrogen 23

CHAPTER 4
Sulfur 35

CHAPTER 5
Phosphorous 41

CHAPTER 6
Calcium 51

CHAPTER 7
Magnesium 69

CHAPTER 8
Potassium 81

CHAPTER 9
Sodium 91

CHAPTER 10
Chlorine 95

CHAPTER 11
Trace Elements 99

CHAPTER 12
Paste Analysis. 123

CHAPTER 13
Soil Balancing Examples 133

CHAPTER 14
Foliar Feeding 173

CHAPTER 15
Soil Nutrients Versus The Weather 179

CHAPTER 16
Modified Growing Media 185

CHAPTER 17
Water Analysis 205

CHAPTER 18
The Future of Agriculture 217

Acronyms, Fertilizer Key 223

Index 225

Chapter 1

The Value of Soil Testing

I WOULD BE REMISS IF I STARTED OUT TALKING ABOUT SOIL BALANCING or soil nutrients without first talking about the correct way to sample soil. Some of you will think to skip to the next chapter, but as the agronomist for Logan Labs in Ohio, I have talked to plenty of people who made some very basic mistakes in soil testing that resulted in wasted time and money.

The real value of soil analysis is being able to make historical comparisons so you can see where the nutrient levels have been and where they are headed in your soils. In order to do this, it is imperative that the soil depth be held constant. A one-inch change in depth for a normal mineral soil represents approximately 330,000 pounds of soil in an acre-furrow slice. An acre-furrow slice is considered to be 6 inches of soil depth over an acre, or 21,780 square feet. The acre-furrow slice represents approximately 2 million pounds. If your soil lab reports your data in parts per million (ppm) and you sampled to a depth of 6 inches, you can multiply the ppm times two and arrive at the pounds per acre of each nutrient. Assuming you sampled to nine inches in depth, the multiplication factor is now three. If you prefer to sample to 7 inches, the factor is 2.33, and so on. This applies to normal mineral soils and not muck soils or modified soil mixes. We will discuss these specific soils in a later chapter.

In the past, this was not quite as important as it is today. Why? Back in the day, when the primary tillage was moldboard plowing, the soil profile was constantly being turned and mixed, so controlling sample depth was not as important. Today, with the implementation of minimum and no-till tillage practices, the stratification of nutrients can be very pronounced, demanding the rigorous control of soil depth. Therefore, in dry years, when the soils are hard, it is easy to go a little shallow—or a little deep in wet years. Variations like these will impact your data coming back and leaving you wondering where you are heading in your soil fertility program.

How do you know where—or how many—samples to take?

I suggest that you keep it as simple as possible when starting a soil sampling program. If your sampling area is small and all one soil type, just take one sample. Although you might be growing a variety of crops and rotating them every year, just take one sample. For larger areas, I prefer to split the soils out according to color or soil type. Sample the light-colored soils separately from the dark-colored soils since they will have different nutrient-holding capacities. Even if you only want to buy one fertilizer mix, by knowing the different soil nutrient levels, you could adjust the spreading of the mix to part or the entire area. Double-spreading the problem areas is something that I commonly recommend. Let's assume you have 10 acres in your cropping area and there is a half-acre area that is different, but you know you are not going to treat it separately. Leave it out of the overall sample. There is nothing wrong with biasing your sample to the problem areas in your high or low ground. You will probably spend a little more money in the long run for fertilizer, but the important thing is to get a sample and start the balancing process. In this day and age, guessing just doesn't cut it. Having too many nutrients is just as bad as not having enough.

I prefer using a soil probe for sampling; however, a spade will work if you are careful and don't have a lot of sampling to do. The spade works best when there is a little moisture in the ground, which helps keep the soil from crumbling off of the spade while collecting your soil sample. Initially, take a spade full of soil and lay it up on the ground close to the hole. Try to go 2-3 inches deeper than the depth you want to sample. Then, with the shovel marked to the desired sampling depth, sheer about a half an inch off the back wall and place it in a clean bucket. Do this four to five times over the desired area. Mix the soil up well and place in a designated soil bag, or allow to air dry and place in a plastic freezer bag. The laboratory generally likes to have around 2 cups of soil. You want to have enough soil so that if the lab would like to recheck something, or you call and want to add an extra test after the results come back, there is enough soil to perform the extra tests.

A soil probe is probably the best tool for soil sampling. It takes a much smaller soil sample than the spade, requiring more areas to be sample, thus increasing the accuracy of the sample. In any field or garden there is one square foot that represents the entire area that you must sample. The problem is knowing where that area is, so the more samples randomly collected the better the data.

When is the best time to sample?

Traditionally, the best time to sample has been after harvest, since the soils would be at the lowest nutrient level after the crop had been removed. This meant the soils could be sampled any time after harvest through winter, and anytime up to planting, just as long as you could get the data back from the lab in time to make the spring fertilizer applications. The number of farmers today has been greatly reduced for a number of reasons; therefore, time is becoming an extremely valuable commodity. Getting a start on next year's growing season in the fall is becoming the norm. The dry fall soil conditions lend themselves more to those farmers using manures, helping to minimize soil compaction; however, this is done at the expense of losing a portion of the nitrogen in the manure. Organic farmers using low-soluble fertilizers like rock phosphate, elemental sulfur, rock dusts, and even limes would benefit from making a fall application even on the lower exchange capacity soils (<10). Unless the fertilizer data is available before harvest, I see some farmers, who are still doing some minimum tillage, forgoing soil sampling in order to get the fall work done. This is less of a problem for the no-till farmers. Therefore, between the time factor and the weather, some soil sampling is shifting from fall to spring and early summer in order to have the fertilizer needs determined and ready to be applied after harvest. Fall applications of the soluble fertilizers containing sulfates or chlorides as the anion lend themselves to being lost over the winter and spring. This is especially critical for the low-exchange-capacity soils. Losing chlorides is not especially concerning, but losing the sulfates when a very high percentage of soils are sulfur deficient is a big concern.

Citing water-quality issues, regulations push for spring applications of fertilizers, but time will tell whether the combination of no-till and big spring rain events will solve the problem.

The bottom line is to try to sample at least forty to sixty days since the last fertilizer application event. This includes manure applications.

Once the samples are collected, the decision as to which soils lab to send them to is very important. If you already have a lab that you are comfortable with, don't change. Bouncing between laboratories looking for the cheapest one will only confuse you. Even though you find two laboratories that are using the same extracting solution—say, Mehlich 3—this doesn't mean there won't be differences in container sizes, shaking times, instrumentation, etc. This will result in some subtle differences in the data.

The goal of soil sampling is to interpret the data back to a response in the field.

If the calcium is low, we expect a certain response in the field-like grass-control problems or poor root development. If potassium falls below a certain level, we might see yellow shoulders on tomatoes or increased disease pressure, etc. All labs have some inherent differences. In this book, when I suggest a balance of calcium/magnesium balance for the base saturation to be 65/15, for the laboratory that you are using it may be 68/12 or 70/12.

You must decide where the best balance is for your crops and soils based on the lab that you are using.

This balance can change based on the crops that you are growing—for example, field corn vs. blueberries.

What tests should you ask the lab to perform on your soil?

I would start with a standard test, which includes total exchange capacity, pH, organic matter, sulfur, phosphorus, calcium, magnesium, potassium, sodium, boron, iron, manganese, copper, zinc, and aluminum. I also like to see the base saturation of cations. I will discuss each one of these parameters in the following chapters. The standard soil test is the test of choice when you know very little about the soil or when the soil is in extremely poor shape. Soils that you have been working on for a while and that have the possibility of growing specialty crops such as vegetables, berries, and fruits are an automatic for paste testing. Even field crops such as corn and beans in high-production environments should have the predominant soil type in each field tested with paste analysis.

The paste analysis parameters should include all the standard soil test parameters, along with chlorides, bicarbonates, and soluble salts. All these parameters and guidelines will be discussed in the next chapter.

There are other tests like available nitrogen and physical testing, but the standard and paste tests are the ideal starting points.

Those who know that their soils are coming from high-pH areas, like the calcareous soils in Texas or the corral-based soils in Florida, might consider asking for an ammonium acetate extraction for the cations. All the other nutrients will be run using a Mehlich extracting solution. These types of soils should all have pH values of 7.4 or higher. The low pH of the Mehlich solution can dissolve free carbonates, causing the total exchange capacity to be exaggerated, resulting in exaggerated desired values. This will be discussed further in the section on total exchange capacity.

How is a soil sample analyzed?

It is important to have some basic knowledge of the testing procedure so you can understand some inherent limits of soil tests. I will attempt to do this without going into an extremely detailed explanation. Basically, all of the soils are air dried when they arrive at the laboratory. Heat drying may cause the clays to collapse, tying up the potassium, which would result in low potassium values on the test.

Once the soils are dried, they are ground and sieved through a 2mm screen. From the grinding room, the samples are moved to the extraction room, where small amounts of soil are collected from the ground soil to test pH, organic matter, and nutrients. In order to analyze for nutrients, 5 ml of soil is volumetrically scooped from each sample and added to an extraction vessel. An extracting solution, which is normally Mehlich 3, is added to the vessel with the soil in it. The samples are placed on a shaker and shook for a specific amount of time. Once the shaking time is complete, the sample is removed from the shaker and filtered through a filter paper and the filtrate is collected in a tube.

The solution is then sent to the ICP (Inductively Coupled Plasma) unit for analysis of all the major and minor elements. The soil that remains on the filter paper is destroyed at this time. As you can see, there are many steps to the get to the final numbers, and not all labs may do the exact same steps. Volumes of soil or solution, shaking time, and even the type of shaking vessel can produce small variations in the results. This one reason why jumping from lab to lab, for whatever reason, can really get you confused when comparing data.

Depending upon the laboratory, different extracting solutions may be used for analyzing the soils. Mehlich 3 is now the industry standard, but some labs may use ammonium acetate or a Morgan extract. Most labs have gotten away from the Morgan extract since it is a sodium-based extracting solution, and if you want to report sodium it is not possible unless another type of extraction is done. The pH of Morgan extracting solution is around 4. This was thought to be the pH in close proximity to the plant roots, since plants exchange hydrogen ions for the cations being taken up the plants. The phosphorus and trace elements required another extracting solution. This slowed turn-around time and increased the labor, which resulted in Morgan being replaced by Mehlich 3. The Mehlich 3 solution will extract all the cations, P-1 phosphorus (the more soluble phosphorus—not rock phosphates) and all the trace elements.

This rapidly became the industry standard because of reduced labor and rapid turn-around time. The pH of the Mehlich solution is approximately 2.5. This can pose an issue on those soils with a lot of free carbonates. The low pH can dissolve free calcium in the soils like those in Florida, which are corral based, or soils in Texas, which have a naturally high calcium concentration. Even fresh lime that has been applied and has not dissolved and attached to the soil exchange sites can be picked up, resulting in higher-than-normal exchange capacities. The lime issues are not normally a problem, but no-till farming increases the risk. In some of the naturally high calcium soils, when pH levels go above 7.4, having the lab extract the cations with ammonium acetate, which has a pH level of 7.0, will reduce the problem of dissolving free calcium, but not completely. Some have gone as far as using an ammonium acetate solution buffered to a pH of 8.2 as the extracting solution for soils with free carbonates. This extracting solution does a great job of not dissolving calcium, but it tends to underestimate the magnesium and potassium (see Lab Sample 1, a high carbonate soil extracted three different ways). None of the extracting solutions are perfect in every situation, but changing the extraction solutions so we can get numbers that we think are more to our liking doesn't make a lot of sense to me.

When selecting a lab, you have three choices. You can get the results fast, cheap, or accurate, but you can only choose two of the three. If you want your results fast and accurate, they probably won't be cheap. If you want the results fast and cheap, they may not be as accurate as you like. I think you get the picture, so choose a lab of your liking and stay there.

			Mehlich	AA	AA 8.2
Sample Location					
Sample ID			3		
Lab Number			4	5	6
Sample Depth in inches			6	6	6
Total Exchange Capacity (M.E.)			36.56	18.47	13.15
pH of Soil Sample			8.3	8.3	8.3
Organic Matter (percent)			3.73	3.73	3.73
ANIONS	SULFUR:	p.p.m.	11	11	11
	Mehlich III Phosphorous:	as (P_2O_5) lbs / acre	57	57	57
EXCHANGEABLE CATIONS	CALCIUM: lbs / acre	Desired Value	9943	5024	3576
		Value Found	13217	6627	4587
		Deficit			
	MAGNESIUM: lbs / acre	Desired Value	1052	532	378
		Value Found	493	267	245
		Deficit	-559	-265	-133
	POTASSIUM: lbs / acre	Desired Value	1140	576	410
		Value Found	184	121	142
		Deficit	-956	-455	-268
	SODIUM:	lbs / acre	43	29	33
BASE SATURATION %	Calcium (60 to 70%)		90.39	89.69	87.21
	Magnesium (10 to 20%)		5.62	6.02	7.76
	Potassium (2 to 5%)		0.65	0.84	1.38
	Sodium (0.5 to 3%)		0.25	0.35	0.55
	Other Bases (Variable)		3.10	3.10	3.10
	Exchangeable Hydrogen (10 to 15%)		0.00	0.00	0.00
TRACE ELEMENTS	Boron (p.p.m.)		0.74	0.74	0.74
	Iron (p.p.m.)		101	101	101
	Manganese (p.p.m.)		132	132	132
	Copper (p.p.m.)		8.61	8.61	8.61
	Zinc (p.p.m.)		9.02	9.02	9.02
	Aluminum (p.p.m.)		94	94	94

Lab Sample 1

Chapter 2

Exchange Capacity, pH and Organic Matter

Let's take a look at a standard soil report (Lab Sample 2) on the following page and discuss each of the parameters as far as what they do and what good starting-point levels would be.

Sample Location
The sample location should mean something to you and it is important to keep it the same from year to year. Logan Labs and I assume others have data archived so you could retrieve historical data for comparisons, but the identification must be identical for the computer to retrieve the data over a specific time period.

Lab Number
This number is generally the number your sample was given for the run of soils for that day. It is a good reference number when calling the lab about a question concerning your soil analysis.

Sample Depth
The sampling depth is very important, as discussed in the previous chapter. It is this number that is divided by three (this assumes every 3 inches of an acre weighs one million pounds) and multiplied by the parts per million (ppm) determined by the Inductively Coupled Plasma (ICP) unit for each of the cations and for phosphorus. This is how the pounds per acre are determined for the phosphorus and cations.

			C1		
Sample Location			C1		
Sample ID					
Lab Number			7		
Sample Depth in inches			6		
Total Exchange Capacity (M.E.)			12.28		
pH of Soil Sample			6.1		
Organic Matter (percent)			5.31		
ANIONS	SULFUR:	p.p.m.	6		
	Mehlich III Phosphorous:	as (P_2O_5) lbs / acre	56		
EXCHANGEABLE CATIONS	CALCIUM: lbs / acre	Desired Value Value Found Deficit	3339 2688 -651		
	MAGNESIUM: lbs / acre	Desired Value Value Found Deficit	353 735 		
	POTASSIUM: lbs / acre	Desired Value Value Found Deficit	383 79 -304		
	SODIUM:	lbs / acre	45		
BASE SATURATION %	Calcium (60 to 70%)		54.73		
	Magnesium (10 to 20%)		24.94		
	Potassium (2 to 5%)		0.82		
	Sodium (0.5 to 3%)		0.80		
	Other Bases (Variable)		5.20		
	Exchangeable Hydrogen (10 to 15%)		13.50		
TRACE ELEMENTS	Boron (p.p.m.)		0.44		
	Iron (p.p.m.)		337		
	Manganese (p.p.m.)		30		
	Copper (p.p.m.)		3.72		
	Zinc (p.p.m.)		2.13		
	Aluminum (p.p.m.)		472		
OTHER	Ammonium (p.p.m.)		1.1		
	Nitrate (p.p.m.)		1.3		

Lab Sample 2

Total Exchange Capacity

Some laboratories report Cation Exchange Capacity (CEC) while others report a Total Exchange Capacity (TEC). The difference is sodium. Those labs reporting a cation exchange capacity are not including sodium in their list of cations. For most soils, sodium is a small portion of the exchange capacity; but for some of the coastal soils and irrigated soils, sodium can dominate a significant portion of the exchange sites on the soil. I will refer to TEC for the remaining discussion in this book.

In theory, we look at the TEC as the cation-holding capacity of the soil. This holding capacity is based on the soil's ability to adsorb cations onto the exchange sites of the clays or humus in the soil. Soils with higher amounts of clay and humus tend to have higher holding capacities and consequently higher TECs. Note the variations in exchange capacities for the following clays and humus.

Type of Clay or Organic Material	Exchange capacity (M.E.)
Kaolinite	3-15
Illite	10-40
Montmorillonite	70-100
Vermiculite	100-150
Humus	50-250

Table 1

The differences in the exchange capacities of the various clays is due to the amount of weathering they have been expose to. The kaolinite clays are typically found in the southern portion of United States. These soils are generally considered older with respect to the geological timeline. It is quite possible to have a soil made up predominately of clay with an exchange capacity similar to sand. I should also point out that both sand and silt have exchange capacities of less than three. Therefore, if you have a soil made up primarily of sand and silt, and you want to boost the exchange capacity, you probably don't want to use clay unless you're interested in concrete. Organic matter or humus would be the product of choice to boost the exchange capacity of any soil. Now that we know what affects the exchange capacity, just how is exchange capacity measured in the laboratory?

Laboratories measure exchange capacity by the summation of cations. Once the ICP data is recorded, the parts per million of each of the cations is converted to pounds per acre by the computer, using the sampling depth that you provided. The pounds per acre of each cation is then converted to milli-equivalents. The following equation is what the computer uses to calculate total exchange capacity.

$$TEC = \frac{(lb./ac\ Ca)}{400} + \frac{lb./ac\ Mg}{240} + \frac{lb./ac\ K}{780} + \frac{lb./ac\ Na)}{460} \times 100$$

$$100 - (\text{other bases} + \text{exchangeable hydrogen})$$

Other Bases		Exchangeable Hydrogen	
= 11.4 - pH	if soil pH > 6.1	= 0	if pH > 7.0
= 17.4 - (2 x pH)	if soil pH > 3.0 & < 6.1	= (7 - pH) x 15	if pH > 6.0 & < 7.0
= 13.3 - (0.6x pH)	if soil pH > 2.2 & < 3.0	= 195 - (30 x pH)	if pH > 5.0 & < 6.0
= 17.4 - (2 x pH)	if soil pH < 2.2	= 145 - (20 x pH)	if pH > 4.0 & < 5.0
		= 105 - (10 x pH)	if pH > 3.0 & < 4.0
		= 93 - (6 x pH)	if pH > 2.2 & < 3.0
		= 155 - (25 x pH)	if pH > 2.2

Table 2

As you can see, calculation has a number of idiosyncrasies as far as pH parameters, but the basic equation is summing the cations to arrive at the TEC. As long as the soils have reached equilibrium from any fertilizer application—meaning the fertilizer has dissolved and the cations are attached to the clay or organic exchange sites—then you will have a true exchange capacity. The problem with exchange-capacity inaccuracy arises primarily with soils containing free carbonates (see Lab Sample 1). This could be naturally occurring free calcium carbonates like corral- or marl-based soils or soils derived from limestone parent material. It could also be a limestone application that is not completely broken down. This is more prevalent in no-till since the applications are surface applied and not incorporated. It is not just a fineness issue for the lime; it is a solubility problem as well. As lime breaks down from a surface application, the pH increases and reduces the solubility of the lime, allowing free carbonates to hang around longer than if the lime was mixed into the soil. When the lime is dissolved by the Mehlich extracting solution, excess calcium is picked up by the ICP and interpreted as an exchangeable cation, which results in an increase in the exchange capacity and the desired values on the test. Lab Sample 1 is a relatively sandy soil, probably with an exchange capacity of less than 10, but the excess calcium has boosted the exchange capacity to over 36. These issues are not limited to just calcium. Any free-floating cation in the soil that is not attached to an exchange site will elevate the exchange capacity and the desired values as well. The actual exchange capacity is something that does not readily change once the soils are in close balance. Really low pH soils that have a lot of exchangeable hydrogen will start out with a higher exchange capacity and drop as you add lime to the soil. This happens because calcium or magnesium is replacing the hydrogen ions on the exchange sites and calcium or magnesium need 400 pounds or 240 pounds, respectively, to equal one milli-equivalent, whereas hydrogen needs only 20 pounds to make one milli-equivalent.

SOIL PH

The soil pH is a measure of hydrogen ions in solution. It measured by an electrode that is placed in a 1:1 soil-to-water mixture. Some labs may also run a buffer pH besides a water pH. The buffer pH is a special solution, that is designed to determine not only hydrogen ions in solution but also the potential hydrogen ions on the colloids that will potentially move into solution once the soils are limed. Soils naturally try to resist or buffer against change with the cations on the exchange sites. Soils high in organic matter and clay tend to resist the change much better than sandy or silty soils.

Some people will say not to worry about the pH and to just balance the cations. It is true that once the cations are balanced, the pH should settle in at around 6.5; however, raising the pH too high even for a short period of time will affect the solubility of major, and especially minor, elements. I have seen this happen on the paste analysis, which we will discuss later. Everyone has seen a pH chart like the one below. When looking at a pH chart, remember that this represents a static situation. For example, when we look at phosphorus at the higher pH range it suggests that phosphorus will be low. That is true if the reason for the high pH is calcium or magnesium, which will precipitate phosphorus out as rock phosphate. There still may be a lot of phosphorus in the soil, but it is not available. If the reason for the high pH is sodium, the phosphorus could still be very available. Look at calcium and magnesium on the lower spectrum of the pH scale. This chart is basically saying that to have a low pH, calcium or magnesium must be low. This is probably true if the pH has been low for some time, but pH levels in the 5.5–6.0 range does not automatically mean that calcium or magnesium availability is low. It is possible for coarser lime particles to exist in the soil and not be totally picked up by the pH test. Rather, the particles will continually dissolve at a low pH, providing nutritional calcium. I have seen paste analysis with a pH range of 5.5–6.0 have much more available calcium and magnesium than soils with pH levels around 7.0. Granted, reserves are rapidly being depleted at the lower pH levels, but availability could still be very good.

Currently I would like to see the pH levels of the higher exchange soils (> 10) be closer to 6.2. I would hold off liming till the pH levels get to 5.8–6.0 or when the calcium solubility starts falling off below 20 ppm. When liming is necessary, keep the amounts down between 2,000 and 3,000 pounds of the appropriate lime, depending upon the balance. In lower exchange-capacity soils I would hold the pH to 6.5 since their buffering capacity is less effective, and I would keep the lime applications to 2,000 pounds or less. I would also limit all lime applications to 2,000 pounds or less even on higher-exchange-capacity soils if the farming method is no-till. Adding too much lime will only raise the surface pH and lower the solubility. It is for these reasons that I really don't see the need for a soil buffer pH, since I don't want to try and fix

How soil pH affects availability of plant nutrients

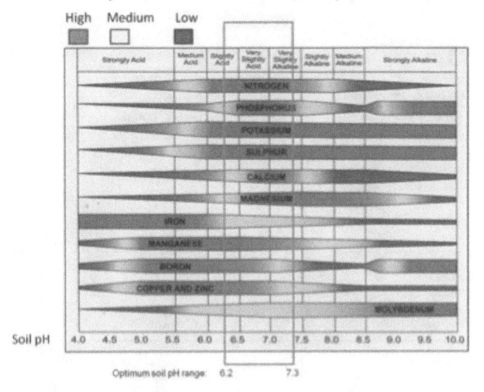

a pH issue with one big lime application.

Another reason for holding the soil pH down in the lower 6 range is that rainwater has been averaging a pH of 6.6. Back in the late '70s and early '80s, the pH of rainwater was running in the mid to lower fives. With the Clean Air Act in place, sulfur has been removed from coal-fired power and industrial stacks with sulfur scrubbers, resulting in an increased pH of the rainwater. This higher-pH rainwater will certainly not help dissolve lime applications held on the surface.

Organic Matter

The organic matter (or OM) is derived by burning a known weight of soil at 360 degrees Celsius and then measuring the weight lost by burning. The reduction in weight is indicative of the percent of organic matter, primarily humus. The temperature is too low to include plant residue or root material. When soils come into the lab, most of the

Water Analysis Report

Job Name	SoilTech	
Contact	Bill McKibben	
Rep		
Submitted By	Bill McKibben	

Company	SoilTech	
Sample ID	304521	
Lab Number	12536	
Run Date	11/9/2017	

Sample Location	Rain	
Sample Name		

Notes

pH		6.6
Hardness	ppm	4.9
Hardness Grains	/gal	0.28
Conductivity	mmhos/cm	0.02
Sodium Adsorbtion Ratio		0.41

		ppm	meq/L	lbs/A in
Calcium	Ca	1.3	0.06	0.29
Magnesium	Mg	0.4	0.03	0.09
Potassium	K	< 0.3	0.00	0.04
Sodium	Na	2.1	0.09	0.48
Iron	Fe	< 0.1		0.01

	meq/L		lbs/A in
Total Alkalinity	3.0		0.68
Carbonate	0.0	0.00	0.00
Bicarbonate	6.0	0.10	1.36
Chloride	2.0	0.06	0.45
Sulfate	2.0	0.04	0.45
Salt Concentration	12.2		2.76
Boron	0.02		
Cation/Anion Ratio		0.99	

pieces of plant residue are removed in the grinding and screening process. Measuring organic matter by the loss from ignition has been used for decades and has been quite dependable until recent years. Over the last five or six years I have been seeing small increases in the organic matter, especially for farms using cover crops. I have also been to meetings where the speakers have bragged of doubling their organic matter levels in as few as four or five years. These statements have bothered me for some time; theoretically, it is virtually impossible to double the humus portion of the soil on a large acreage scale in such a short period of time.

Let's consider taking a 2 percent organic matter (humus) soil and raising it to 4 percent. In an acre-furrow slice, which we assume to weigh 2 million pounds, this would mean that a 2 percent organic matter soil would have 40,000 pounds of organic matter or humus. In order to double that level, we would have to add another 40,000 pounds of humus. These 40,000 pounds must be humus and not residue. Literature tells us that composting residue to humus may be as high as 100:1, and present-day composting facilities will tell you that a lot of material will become a small amount of compost, which is more than likely not completely humus. Humus is considered the final product in the composting process when the microbes can no longer degrade the organic residue. To transform a 2 percent OM to a 4 percent OM soil, approximately 4 million pounds of residue would have to be added to an acre to make this happen. That is 2,000 tons of material.

So why are the OM levels rising on the soil test? I work with a refinery operation that uses bacteria to digest suspended oil in their water treatment facility. When the bacteria die, they drop to the bottom of the tank, where they are pulled off and belt-pressed into a cake, which is land-farmed. I had the dead bacteria, also known as biosolids, analyzed as a soil for organic matter, and low and behold the test came back 60 percent organic matter. This indicates to me that those rapid increases in soil OM levels are probably due to the test picking up bacteria, mycorrhizae, fungi and who knows what. These increases are more than likely not humus but an unstable source of organic micro-organisms. Some the fastest increases in OM that I have heard about come from people putting their fields into sod for cattle. Besides picking up micro-organisms in the OM test, it is possible that some sampling variations may be occurring. In developing a dense matt of grass, roots will penetrate the soil; but where does the soil go in order to make room the grass roots? The soil has to actually fluff up to accommodate the roots. Soil granulation will also take place very quickly. This is a significant change in bulk density and a reduction in the amount of soil that would have normally been sampled had a root mass not developed. In addition, sample depth could also be affected by the fluffing up of the soils, assuming the depth is controlled from the surface. This would mean that samples may be collected at a more shallow depth than where they were originally collected, resulting in an increase in the OM.

The bottom line is that increasing organic matter is a very good thing, but the benefits that you get from increasing the humus side of the OM is much higher. Things such as an increased nutrient-holding capacity, water holding capacity and buffering capacity are a few of the benefits from the humus side of the OM. Regardless, if an increase on the organic matter test is from microbes, fungi, or who knows what, it is a positive sign of a healthy and increasingly productive soil.

Organic matter will release nitrogen over time. The following table is a way to calculate nitrogen release during a growing season. This chart is based on the humus portion in the soil. If a portion of the OM is the result of soil biology, the chart might be on the conservative side and the release could be faster. These charts normally considered the nitrogen to be released over the entire growing season. In an analysis of the biosolids that I mentioned earlier, the carbon-to-nitrogen ratio was 8.7:1 with a nitrogen level of 3.2 percent. Humus is generally considered to have an approximately 10:1 C/N ratio.

Estimated Nitrogen Release from Organic Matter (Humus)

ENR=	20 + [(O.M. - 0.5) x 40]	if O.M. < 1%	
ENR=	40 + [(O.M. – 1.0) x 20]	if O.M. > 1%	and < 3%
ENR=	80 + [(O.M. – 3.0) x 10]	if O.M. > 3%	and < 5%
ENR=	100 + [(O.M. – 5.0) x 5]	if O.M. > 5%	and < 10%
ENR=	125 + [(O.M. - 10.0) x 5]	if O.M. > 10%	and < 20%
ENR=	> 130 lbs. of N	if O.M. > 20%	

Table 4: *Estimated Nitrogen Release from Organic Matter (Humus)*

Chapter 3

Nitrogen

> - **Ideal soil level:** Crop and soil dependent.
> - **Mobile in the plant:** Yes
> - **Xylem and phloem mobile:** Yes
> - **Site of initial deficiency symptoms:** Older leaves
> - **Role in the plant:** Formation of amino acids, vitamins and cell division. Major component of plant proteins.
> - **Deficiency symptoms:** Yellowing at the base of the plants and moving up as the deficiency grows. In grasses, the yellowing will start at the base of the plants, with a yellowing starting at the tip and moving down the mid-rib of the leaves. Plants' growth will slow substantially and they will eventually completely yellow.
> - **Toxicity symptoms:** Very dark green, excess vegetative growth, reduced flowering, lodging, and increased disease and insect pressure.

AFTER REVIEWING ESTIMATED NITROGEN RELEASE FROM ORGANIC matter, discussing nitrogen as a fertilizer is a natural follow-up. Nitrogen fertilizer is looked upon as both a villain and the good guy. Nitrogen can consume soil carbon, leach and pollute water sources, increase disease, and help deplete calcium in the soil; on the other hand, it is the one nutrient that drives the boat when it comes to plant growth. If nitrogen goes short, yields and quality suffer substantially, as does the uptake of many of the other nutrients. Nitrogen is one the most difficult nutrients for me to consult on. Every crop has a different requirement level, and the same crop has different levels based on yield. Soil type, organic matter levels, types of cover crops, drainage, and weather conditions are just a few factors that will impact the nitrogen requirements and availability for a crop. It is not good enough to have the nitrogen in the soil at planting; it must remain available through the entire vegetative growth stage and even into the reproductive stage, but generally at a lower rate.

Nitrogen is relatively unstable and subject to loss in all its phases. Nitrogen in the form of plant protein is initially the most stable form, but once incorporated into the soil, the soil microbes begin to break it down into urea and then into ammonia and eventually into the most vulnerable phase, nitrate nitrogen. During this breakdown phase, ammonia toxicity to plant roots and seed germination is quite possible. I have seen a 60–70 percent reduction in seed germination when a lush rye cover crop was worked into the soil and planted a few days later.

Urea (CH_4N_2O), which is a carbon-based protein, has no charge and cannot attach to the soil colloids. Therefore, it is subject to being washed from the soil during a big rain event. Soil microbes can quickly dismantle the carbon-based nitrogen into ammonia, but high-pH soils will slow down the rollover of ammonia (NH_3), which is volatile, to ammonium (NH_4), increasing the risk of volatilization. Ammonium (NH_4) carries a positive charge and can attach to the soil colloids. Therefore, surface-applying urea or urea-based manures like chicken or turkey manure should be incorporated on high pH (>6.8) soils or even lower pH soils if recently limed.

Nitrate nitrogen is the end result of the nitrogen degradation cycle. Plants will absorb nitrate or ammonium, but the plants must expend more energy pulling in nitrate nitrogen. Once inside the roots, the nitrogen is converted back to urea/proteins. Nitrate nitrogen carries a negative charge and basically free floats in the soil solution and can be carried out of the root zone with excessive water. Saturated soils result in a low oxygen situation and soil microbes will pull off oxygen molecules from the nitrate molecules, resulting in a volatile nitrous oxide and off-gassing into the atmosphere.

As you can see, nitrogen management is a delicate balance between soil chemistry, soil physics, soil biology, and weather. No other nutrient is quite as affected by these variables as nitrogen. How do you farm to hold nitrogen in the soil so it is available when the crop needs it?

Multiple applications are by far the best way to reduce nitrogen loss and increase plant availability. Planting cover crops to tie up nitrogen as well as generate nitrogen from legumes is another potential way to hold nitrogen in the soil. Depending where you are in the country and when you finish harvest in the fall, the length of growing time between seeding a cover crop and planting a cash crop may be too short to accumulate enough nitrogen for a cover crop to be profitable on a nitrogen basis. Switching to a cereal rye when it gets late may the best alternative.

So how should one estimate the nitrogen need for a crop? First you need to have some idea of the potential yield of the crop you are going to grow and the potential removal rate (see the partial crop listing below). The NPK values for some of the crops indicate the grain or fruit removal versus the stover or vines. Take corn as an example. If you are harvesting just the grain, the removal is first value, but if you were taking off corn silage, the removal would be both the grain and the stover. Nevertheless, the nutrient level to grow a crop must include both the plant and the grain or fruit.

Nitrogen, Phosphorus, and Potassium Crop Removal
Pounds per crop unit.

Crop	Yield/ac.		N	P in P$_2$O$_5$	K in K$_2$O
Field Corn	1.0 bu.	Grain/ Stover	0.9/0.45	0.38/0.16	0.27/1.10
Soybeans	1.0 bu.	Grain/ Stover	3.80/1.10	0.84/0.24	1.30/1.00
Wheat	1.0 bu.	Grain/ Stover	1.50/0.70	0.60/0.16	0.34/1.20
Tomatoes	40 ton	Fruit/Vines	144/88	67/20	288/175
Peppers	12 ton	Fruit	137	52	217
Sweet Corn	5.0 ton	Corn/Stalks	55/100	8/12	30/75
Potatoes	25 ton	Tubers/Vines	160/102	80/34	264/90
Cabbage	20 ton		125	30	130
Cucumbers	10 ton	Fruit/Vines	40/50	14/14	66/108
Peas	3 ton	Peas/Pod & Vine	45/100	9/17	17/62
Onions	20 ton	Bulb /Top	110/35	20/5	110/45
Muskmelon	11 ton	Fruit/Vine	95/60	17/8.0	120/30
Table Beets	25 ton	Root/Tops	170/190	30/13	210/370
Lettuce	20 ton		90	30	185
Carrots	25 ton	Roots/Top	80/65	20/5.0	200/145
Apples	12.5 ton	Fruit/New Wood	20.0/80.0	8.6/38.0	50/130
Peaches	600 bu.		95	40	120

Table 5

How does this work for a growing crop? Let's just pick out tomatoes in which we are expecting to grow 40 tons/ac. Checking the chart, we see that we are going to need approximately 144lb/ac of nitrogen for the fruit and 88lb for the vines, for a total of 232lb/ac of total nitrogen. If you typically only grow 30 ton/ac, you would need 75 percent of 232, or 174lb, of total N. We will discuss the phosphorus and potassium in each of their sections. To grow 40 tons/ac, we need 232lb of nitrogen either from the soil, cover crops, manure, or manually applied nitrogen fertilizer. Assuming that we are growing on an 8.0 percent organic matter soil, we can calculate the estimated nitrogen release (ENR) from the chart on page 26. An 8.0 percent organic matter level in the soil, based on the estimated nitrogen release table, will release approximately 115 pounds per of nitrogen. The following is how the calculation is done.

ENR= 100 + (8.0 (our OM) – 5.0) x 5 if O.M. > 5 percent and < 10 percent. This equals 100 + 15, or 115 pounds of nitrogen that will be generated from an 8 percent organic matter level.

If we need 232 pounds of total nitrogen and the OM is generating 115, I still need to supplement 117 pounds of nitrogen. If I happen to have planted a legume cover crop last fall after the previous crop, I could subtract more nitrogen from the remaining 117 pounds that I need. You should check with your seed supplier or local extension as to the amount of nitrogen you might expect from your cover crop. If the cover crop could generate a consistent 40 pounds of nitrogen, then just under 80 pounds of nitrogen would need to be supplemented from either a commercial or organic source. Keep in mind that if you are carrying over a large amount of residue from the previous crop, that might impact the amount of available nitrogen for our tomatoes, or possibly cause a delay in the nitrogen release time. Soil micro-organisms will almost always get first chance at the soil nitrogen for breaking down residue in the soil. The wider the carbon-to-nitrogen ratio of the existing residue, the longer or more nitrogen will be needed to satisfy the micro-organisms. Note the chart below.

Carbon to Nitrogen ratios of various crops.

Crop	C:N Ratio	Impact on Nitrogen
Wheat	80:1	Reduce N availability
Corn Stocks	55:1	Reduce N availability
Mature Cereal Rye Cover Fall Seeded	40:1	Reduce N availability
Immature Cereal Rye Cover Fall Seeded	25:1	Neutral effect
Crimson Clover Fall Seeded	21:1	20-30lb
Annual Ryegrass Late Summer Seeded	20:1	75lb
Red Clover Mature, Seeded in Wheat	15:1	75-80lb
Hairy Vetch, Seeded after Wheat	11:1	100-120lb
Microbes	8:1	?

Table 6

The profitability of legume cover crops following soybeans going to corn is hard to pencil out just on a nitrogen basis. Late summer planting following wheat or vegetables going to corn is certainly more profitable. To get the high nitrogen values, these legumes need a longer growing period than you would get following a soybean crop.

There are some nitrogen monitoring tools available that can be employed on field corn that can help to maximize nitrogen efficiency. The first is a soil nitrogen test. This test monitors ammonium and nitrate nitrogen in the soil. This test is mainly used as a pre-side dress test to determine the nitrate and ammonium nitrogen level in the soil from manure applications. Knowing the nitrogen in the soil will help determine the rate of side-dress nitrogen. This test does not pick up urea-based proteins from unbroken-down manure or cover crops or urea-based fertilizers. This test may also be used as monitor to verify how well the corn crop utilized the side dress nitrogen. By probing the side dress trench left in between the rows, you can see if there is any unused nitrogen left. Background nitrogen is between 5 and 10 ppm, depending upon the organic matter, so if there is much more nitrogen than this, not all the nitrogen was used up by the crop. Be careful how you interpret the numbers that you get back from the lab when sampling band applications. Let's say you get a total of 40 ppm of nitrate and ammonium from the lab and the sample depth in the side-dress trench was 12 inches. If every 3 inches represents one million pounds, a sample depth of 12 inches represents 4 million pounds, resulting in a 40 ppm test being translated into 160 pounds per acre of excess nitrogen. However, the level of nitrogen in parts per million does not represent the whole area, since the nitrogen was banded. I generally consider the affected band of soil to be about 6 inches at the most. Therefore, if the nitrogen was banded every 30 inches, the amount of nitrogen on the report represents one fifth of the total, or 32 pounds of total nitrogen. Secondly, 90 to 95 percent should be in the nitrate form, which indicates there was sufficient amount of aeration and biological activity to convert the urea-based or ammonia-side-dress products.

Another test to go along with the soil nitrogen analysis is the corn stalk analysis. Even though you're checking side-dress band and the corn stalk analysis after the season is over, they can help make adjustments for the following year. I had a client who really liked to do a fall nitrogen application for his corn crop the following year. Instead of applying nitrogen in the spring, he could be planting his corn and be ahead of the game. His yields were OK, but I thought they could be better. After using the stock nitrogen test three years in a row and seeing that the nitrogen levels in the stocks were unacceptably low, he switched to spring application and yields jumped 20-30 bushels for the same amount of nitrogen that he was using. Many farmers in the west are fall-applying nitrogen and more than likely they are like my client, who was losing a portion of his nitrogen before the season. Unfortunately, much of this nitrogen ends up in the Mississippi River and eventually in the Gulf of Mexico, creating a huge dead zone. To do the stalk test correctly, you must come up off the ground 6 inches and cut the next 8 inches of the stalk for analysis. Six or eight representative stalks is all that is needed per field in order to determine stalk nitrogen levels. This must be done after physiological maturity of the corn, which is commonly known as black layer. This is the point in which no more carbohydrates or protein is entering the kernel of corn. The corn will be roughly 37 percent moisture at this time. Taking a stalk test before black layer will normally result in excessive nitrogen levels unless the plant is visually nitrogen deficient above the ear. Just because you see visual nitrogen deficiencies on the lowest leaves of the corn plant does not mean more nitrogen should have been applied to the crop. Ideal numbers are between 750 and 1,000 ppm. Numbers below 750 indicate that not enough nitrogen was available to the plant. There is data to suggest that nitrogen values down to 500 may mean a 2- or 3-bushel reduction in yield, but below 500 it is anyone's guess what the yield loss may be. We were seeing levels in the 100-150 range for my client who was applying nitrogen in the fall. Nitrogen values above 1,000 indicate that there was more nitrogen available for additional yield. Calculations and field results have shown that for every 1,000 ppm above the desired 1,000 ppm, ten pounds of excess nitrogen was applied or there was enough nitrogen in the plant to produce an additional 12.5 bushels of corn. Before I cut nitrogen, I try to figure out what I might do to increase the yield. The easiest thing to try is increasing the population at planting. The bulk of the nitrogen used in the U.S. is for corn production. These simple tests need to be used to prevent the contamination of ground water and surface water.

Excess nitrogen in vegetables show up as too much vegetative growth and lodging. There may also be a reduction in flowering for things like tomatoes and peppers. Some tomato varieties may show fruit cracking and cabbage heads that split open, especially after a warm summer rain. Greens tend to be bitter since ammonium nitrogen may compete for magnesium and potassium uptake. Increase disease is a real threat in the vegetable business. Excess nitrogen results in rapid growth, causing the cell walls to be soft and thin, making them vulnerable to insect and disease infestations.

Microbes will try to maintain a carbon-to-nitrogen balance of around 10:1, so excess nitrogen will be the fuel for the breakdown of carbon residue or even organic matter.

Nitrogen is very mobile in the plant, so deficiencies show up on the older portions of the plant first, but as the deficiency progresses the yellowing will move up the plant to the new growth. Since nitrogen shortages can be very devastating to yield and quality, tissue sampling can be used to head off this problem. Typically, tissue sampling calls for the first mature leaves to be sampled, but if the deficiency is exhibited on the old growth of the plant first, it will be too late when you catch it using the normal procedure. Sampling the first mature new leaf and the older leaves at the bottom of the plant (not the ones shaded out and dying at the very bottom) is the best way to head off mobile nutrient deficiencies. This applies not only to nitrogen, but phosphorus, potassium, magnesium, molybdenum, and nickel.

Sources of nitrogen are many. Here is a partial list of common fertilizers with nutrient concentrations.

Sources of Nitrogen

Product	Type	Effect on pH	% N	% P_2O_5	% K_2O
Anhydrous Ammonium	Gas	Acidic	82	0	0
Urea	Pellet	Acidic	46	0	0
Aqua-Ammonia	Liquid	Acidic	28-30	0	0
Ammonium Nitrate	Granular	Acidic	33	0	0
Ammonium Sulfate	Granular	Acidic	21	0	0
Di-ammonium Phosphate	Pellet	Acidic	18	46	0
Mono-Ammonium Phosphate	Pellet	Acidic	11	52	0
Calcium Nitrate	Pellet	Alkaline	15	0	0
Sodium Nitrate	Granular	Alkaline	16	0	0
Potassium Nitrate	Granular	Alkaline	14	0	37
Fish Meal	Meal	Acidic	10	8.2	0.5
Blood Meal	Meal	Acidic	13	0.6	0.4
Corn Gluten Meal	Meal	Acidic	7	1.0	0.3
Feather Meal	Meal	Acidic	13	1.4	1.8
Crab Meal	Meal	Acidic	5	3.6	0.5
Alfalfa Meal	Meal	Neutral	3	0.7	2.9
Compost- Variable Sources	Meal	Neutral	1-4	0.5-6.9	0.6-3.6

Table 7

As you can see, more than just nitrogen may come along with your product of choice. If phosphorus is already high in your soil, ammonium phosphates, or fish meal, may not be the best choice.

Here is a list of common manures and their fertilizer concentrations.

Product	% Dry Matter	N /ton	P_2O_5 /ton	K_2O /ton
Dairy Manure	18%	9	4	10
Horse Manure	46%	14	4	14
Chicken Manure	45% w/o litter	33	48	34
Chicken Manure	75% w/ litter	56	45	34
Turkey Manure	29% w/o litter	20	16	13
Swine Manure	18% no bedding	10	9	3
Beef Manure	15% on concrete	11	7	10

Table 8

There are many variations in manure values, depending upon how the manure was handled, such as with or without bedding. Manure exposed to rainwater, stored in a lagoon, or composted will either dilute or concentrate the nutrients. A manure analysis is the best way to know for sure what the manure contains. All the manure samples in table 8 and the protein sources in table 7 come with other nutrients. Balancing soils just using manures generally creates an imbalance of either phosphorus or potassium.

Applying Nitrogen Correctly

Applying nitrogen to grow a crop is not just about putting the right amount of nitrogen on ahead of or during the growing cycle. Rather it is having the right amount of nitrogen available to the plant when it needs it. Knowing the demand time will allow you to pick and choose the kind of nitrogen and the best time of application. Let's compare growing corn versus beans. Corn will need a lot of nitrogen from about knee high right through to kernel set. The bulk of the nitrogen is used for the plant growth. When you look at the nitrogen needed to produce the protein in the kernels for 200 bushels of corn, it is roughly only 125 pounds of the 270 pounds of nitrogen needed for the total crop. Beans, on the other hand, need the bulk of their nitrogen rather late in the season to finish the high-protein seeds. Just the beans in a 70-bushel crop will need nearly 250 pounds of the nearly 350 pounds of nitrogen needed for the total crop. Fortunately, legumes produce a lot of nitrogen through their nodules, but nodule activity begins to drop off once pods begin to set. Deficiencies in molybdenum and copper will affect nitrogen production in the nodules and in the utilization in the plant. The odds are really good that a yield of 70-plus bushels of beans will need supplemental nitrogen applications late in the season.

Vegetables need the bulk of their nitrogen early in the crop cycle and we want to be running low as fruit begins to develop. Too much nitrogen remaining late in the season can lead to cabbage heads splitting, broccoli and cauliflower bolting, and tomatoes cracking. Diseases will be elevated due to soft growth in the plant and fruit.

Nitrogen management is as much an art as it is a science, especially given variable weather patterns from year to year.

Chapter 4
Sulfur

- **Ideal soil level:** 25-35 ppm
- **Mobile in the plant**: Minimal
- **Xylem mobile:** Primarily
- **Site of initial deficiency symptoms:** New growth
- **Role in the plant:** Formation of amino acids, enzymes, and vitamins. Sulfur will impact seed development, disease tolerance, level of nitrates in plants and chlorophyll production.
- **Deficiency symptoms:** Plants will be pale green to yellow if the deficiency is bad enough. The new growth will show the deficiency first, unlike nitrogen, which starts on the lower/older growth. The young, deficient leaves may also exhibit inter-venial chlorosis. A sulfur shortage also causes plants to be shorter and squattier than a normal plant. The yellow leaves may also be more upright and narrower.
- **Toxicity symptoms:** Sulfur toxicity is rare, but sulfate burn caused by the foliar feeding of too many sulfate-based nutrients can happen quite easily. Start by using less than 4-5 pounds of the combination, or of a single sulfated nutrient, per one hundred gallons.

SULFUR IS A PART OF THE ENZYMES IN THE PLANT THAT HELP IN THE nitrogen reductase process. This process allows nitrate nitrogen to be converted into amino acids, thus reducing the build up of nitrates in the plants. Excess nitrates can lead to disease and pest issues along with excessive growth.

The bacteria in legume nodules need sulfur to help fix atmospheric nitrogen. A sulfur deficiency could lead to a nitrogen deficiency in legumes—which if used for feed will be low in protein—and as a cover crop, a reduction in nitrogen for the following crop.

Plants deficient in sulfur also have a significant reduction in chlorophyll and they consequently lower the carbohydrates or energy in the plant. For crops like corn, grown specifically as an energy source, this may mean lower test weight.

Plants can store sulfur as sulfates in the leaves and stems for remobilization at a later time. Bulbous plants like radishes and onions—along with cabbage, alfalfa, and corn—have a high affinity for sulfur. This is one of the reasons why cabbage, onions, and radish cover crops give off hydrogen sulfide and smell so bad when they begin decomposing in the field.

The ideal nitrogen-to-sulfur ratio in plants is around 10:1. The high-sulfur-affinity plants may have a little lower ratio, but plants such as cereal grains may reach 15:1. Younger plants generally have a higher ratio and the ratio tends to drop as the plants mature.

A good rule of thumb: as you increase nitrogen in your fertilization program, proportionately increase your sulfur. Strive for soil levels of nitrogen to sulfur to be around 5:1.

Sources of Sulfur

Types of Sulfur	Effect on pH	% Sulfur
Elemental Sulfur	Acidifying	90.0
Ammonium Thiosulfate	Acidifying	26.0
Ammonium Sulfate 21-0-0-24	Acidifying	24.0
Potassium Magnesium Sulfate (KMag)	Neutral	22.0
Potassium Sulfate 0-0-50	Neutral	18.0
Magnesium Sulfate (Epsom Salts)	Neutral	13.0
Calcium Sulfate (Gypsum)	Neutral	12.0

Table 9

These are the basic minerals with high concentrations of sulfur from which many various combinations of sulfur-containing compounds would be made. There are also other sulfated minerals such as copper sulfate, zinc sulfate, etc., but these would be used in such small quantities that they would have very little influence on the overall concentration of sulfur in the soil.

General Comments on Sulfur

The soils that I work with, along with those that come from Logan Labs for recommendations, are easily running over 90 percent deficient from the 25-35 ppm that I would like to see. This is a far cry from what I use to see just a decade ago. The Clean Air Act is primarily the reason for the significant drop in soil-available sulfur. It is not worth debating whether the Clean Air Act was the right thing to do, but it does have some far-reaching consequences. Plants grown in sulfur-deficient soils will be vitamin-A deficient, are quite possibly more susceptible to diseases and insect damage from nitrate accumulation, and possibly contain lower protein and carbohydrate levels.

Where's the proof? Over the last decade, the use of fungicides for agricultural crops has mushroomed. It is now considered a no-brainer to use a fungicide on soybeans for about a 5-bushel yield advantage. Corn fungicides are rapidly on the rise and many elevators will guarantee acceptance of a farmer's wheat if they used a fungicide. Nearly all high-end sports fields and golf courses use fungicide to one degree or another. For the most part I would agree that fungicides have helped maintain high yields and turf quality. They are promoted as products that increase yield and plant health, when in fact they really prevent yield loss and disease. I truly believe that we are merely treating the symptom and not the real problem of poor nutrition. This issue will not be resolved with only a standard soil report, but with the use of paste testing and tissue analysis. Both of the latter tests will be discussed in detail later.

Who knows about a vitamin A deficiency, since farmers both organic and commercial are paid for quantity, not quality, at this point.

Sulfur levels should run between 25 and 35 ppm during the growing season. One might stay to the lower limit for cereal grains and most vegetables, but use the higher level for the high-affinity crops like cabbage, bulbous crops, corn, and alfalfa. Other than elemental sulfur, all the major sources of sulfur are sulfates. Like nitrates, they can be leached out of the root zone with excessive water moving through the profile, but unlike nitrates they are not denitrified and tend to accumulate in saturated soils. Soil tests showing high sulfur levels without receiving sulfur applications should be suspect for poor drainage. How high is high? If the bulk of your soil tests show background numbers to be between 6 and 10 ppm and one or a few of the tests come back with 20-30 ppm or more, a red flag should go up concerning internal drainage.

It is interesting that what appear to be background levels in the soil for sulfates are also very close to what I see for nitrates. Organic matter and biological activity will certainly impact the background levels of both sulfates and nitrogen in the soil. The best time to apply sulfate forms of fertilizer is just ahead of planting. This is rather difficult for bulk sources of gypsum due to timing and compaction issues, but the later in the fall the better. Sulfate forms such as gypsum and potassium sulfate are pH neutral, but they will tend to lower magnesium over time. Once the magnesium gets to 15 percent, it would be wise to include magnesium sulfate and KMag as part of your sulfur program.

Correcting Sulfur Issues

What is the best way to determine the amount of sulfur products to use?

First start with the standard test soil data. Let's assume the test comes back at what you might consider as background levels—10 ppm. I have indicated that 25-35 ppm are good starting targets. Sampling to a depth of 6 inches, this means that we need between 50 and 70 pounds per acre in the soil (6 inches = 2,000,000 lbs. x 25 ppm = 50 lbs.). This assumes that the plants can pick up the background sulfur and that it will be replaced over time. This is more likely with anions like sulfate, but it is generally not the case for cations. For example, potassium cannot be pulled down to zero on the soil test by plant adsorption. If your soil has 10 ppm then you have 20 pounds available and you would need 30 to 50 pounds of sulfur, depending on the crop.

Depending on the overall soil balance, you could go to the above list and choose the ideal product. If calcium was low, bulk or pelleted gypsum might be a good choice; if potassium was low, potassium sulfate might be the product of choice. Even if calcium is okay on the standard soil report but magnesium is high, gypsum could still be a good choice when trying to use the Albrecht philosophy to balance the calcium and magnesium on 10-or-greater TEC soils. The concept is to replace magnesium on the colloid with calcium, leaving magnesium in solution with sulfate ions, essentially forming Epsom salts, which are highly soluble and capable of leaching out of the soil profile, providing there are no compaction layers and adequate internal drainage. This will be covered again in the calcium and magnesium section.

I reserve elemental sulfur for adjusting soil pH and generally I do not use it as a sulfur source for a cropping program. The degradation of elemental sulfur is based on the soil biology, so it is hard to predict the rate of break down and whether it will be fast enough to meet the crop demand. If forced to use it, I would fall-apply and work in. In order to prevent an acid band at the surface of the soil, I don't like using too much elemental sulfur at any one time on turf or no till. Elemental sulfur should be left out of modified soil mixes that are commonly used in potting soils or greenhouse mixes unless the *Thiobacillus* bacteria necessary to degrade elemental sulfur are present.

Chapter 5

Phosphorus

- **Ideal soil level:** 250 lb./ac. P_2O_5, 110 lb./ac. P, 55 ppm P for general crops. 500 lb./ac. P_2O_5, 220 lb./ac. P, 110 ppm P for specialty crops (vegetables, alfalfa, cannabis, etc.)
- **Mobile in the plant:** Yes
- **Xylem and phloem mobile:** Yes
- **Site of initial deficiency symptoms:** Overall size of the plant. Usually occurring early in the season.
- **Role in the plant:** Regulates the energy in the plant, cell growth, root, and seed formation; promotes winter hardiness; necessary for carbohydrate and protein synthesis; promotes lignification.
- **Deficiency symptoms:** Phosphorus deficiencies mainly inhibit plant growth, resulting in a smaller, darker plant with reddish coloration. The new leaves will be smaller than normal. A phosphorus deficiency will also affect a plant's reproductive performance with fewer and smaller seeds.
- **Toxicity symptoms:** Excess phosphorus rarely leads to a terminal toxic effect on plants. Excessive levels can exacerbate a deficiency in calcium, boron, iron, manganese, copper, and zinc. Zinc and iron would normally be the two nutrients most affected. Older tissues of plants will become excessive if the whole leaf tissues exceed 1 percent. Things like little leaf in the new growth or premature ripening may occur due to zinc tie up. If iron becomes substantially affected, plants may exhibit chlorosis of the tissue.

General Discussion

Phosphorus is like the most popular kids in school. Someone is always hanging around with them. Phosphorus is so popular with other nutrients that it can get tied up very easily. Calcium and magnesium will tie up phosphorus into a form of rock phosphate, especially at higher pH levels (above 7.0). Lowering the pH level below 6.5, calcium phosphates can solubilize and become available again. However, at the lower pH levels (below 5.5) iron and aluminum are extremely efficient at tying up phosphorus into a very insoluble complex. Raising the pH may not help solubilize the phosphorus for quite some time. Phosphorus availability is highly influenced by microbial populations as well as the pH level. I have evaluated tests performed by a company in California that adds micro-organisms to the soil to improve the overall nutrient availability, and the only nutrient consistently improved was phosphorus. Phosphorus is so well liked by all the positively charged particles in the soil that once it is applied to the surface of the soil, it will barely move more than an inch in the soil profile. This causes stratification of the phosphorus that will be discussed at the end of this section.

Phosphorus is a unique anion, and, like nitrogen, it cannot be replaced to any degree by another anion. Phosphorus is only stored in the plant in the form of phytin. Phytin is a calcium-magnesium salt of phytic acid, which is found in tubers and seeds and acts as the storage reserve. Once there, however, it is not available back to the plant. Phosphorus is constantly being recycled in the plant. This lack of storage is why small, rapidly growing plants require a higher percentage of phosphorus in comparison to more mature plants. Phosphorus is constantly being taken up during the growing season, with more than half of the total phosphorus needed for the crop being taken up after flowering.

The mobility of phosphorus is highest in the soil within the pH range of 6.0 to 6.5.

It is also highest when the soil is at field capacity, well aerated, and warm, and when microbial populations are at their peak. Dry soils severely restrict phosphorus availability due to the drop in microbial activity and the increased concentration of cations in the soil solution, resulting in soluble phosphorus precipitating out.

Ammonium (NH_4) ions increase the uptake of phosphorus and nitrate (NO_3) conversely decreases phosphorus uptake. Using ammonium sulfate as a nitrogen source would be beneficial if the P levels were on the low side and the soils were cold.

Since roots practically have to seek out the phosphorus, anything reducing root mass will have a real impact on the P levels in the plant. Compacted soils, anaerobic conditions and root-feeding insects can lead to plants being phosphorus deficient even though there may be adequate levels in the soil. Selecting crop varieties with a more aggressive root system will help in these situations.

Phosphorus-deficient plants tend to look darker in color and more compact. The darker color is the result of the reduced growth of the plant, but the chlorophyll production is not affected. Deficient plants also have an impaired cellulose production, leading to lodging issues. Plants suffering from low phosphorus have trouble converting sugars to starch, which leads to an accumulation of sugars and the production of anthocyans, resulting in a red coloration. Severely deficient plants will stop sugar production all together since phosphorus is critical for the plants to convert sunlight energy to chemical energy (sugars). Plants suffering from phosphorus deficiency later in the cropping cycle may exhibit dark spots on the lower leaves as a result of mobilization to the newer leaves for grain fill. This is more prevalent on beans, beets, potatoes, and clovers. Tillering of small grain plants may be reduced under low phosphorus levels. This might happen even with good soil P levels if the soils were cold. Keeping a little phosphorus in the spring top dress may be necessary even though you are treating the symptom and not the problem.

Phosphorus stratification and the environment

The issue of phosphorus stratification in the soil, as it relates to minimum and no-till tillage, has a significant potential to cause environmental concerns when it comes to water quality. Currently many agencies are trying to apply a bandage on the problem through cover cropping and phosphorus application regulations, but eventually they will have to meet the problem head on. The fact of the matter is that surface-applied phosphorus will not migrate into the soil and remains at the soil surface. I have tested the soils for one of Ohio's No-Till Farmer of the Year recipients, who at the time had been no-till farming for over twenty years, and found through 1-inch sampling increments that 66 percent of all the phosphorus of a 9-inch sample was in the top 3 inches. Does this create a nutrient problem for crops? Not as long as there is root activity and moisture at the surface. The phosphorus mobility listed above says it is both xylem and phloem mobile. Therefore, as long as the phosphorus is available to the roots, it can be moved either up to the stems and leaves or down to the roots below the zone of accumulation. As far as the plants go, they could care less where it comes from in the soil, but the problem arises when dry weather hits and the surface dries out and root activity in the surface declines. Phosphorus availability is substantially depressed in dry-soil conditions, leading to phosphorus deficiencies if the deeper roots can't supply the demand.

Two major things happen when soils begin to dry, and both are detrimental to phosphorus solubility. First, soil microbes rapidly begin to decline, and secondly, the soil solution increases in cation concentrations, precipitating out available phosphorus. This is what happens to plants and phosphorus when a drought event strikes; but what happens under excess rainfall events? Assuming that this situation is created due to no-till and the band-aid of cover cropping is being practiced, what happens to our concentrated phosphorus now? We know that cover cropping is excellent for holding soil from erosion and that cover cropping promotes biological micro-organisms along with nightcrawler activity. These are all positive things for a healthy soil. The problem as I see it with phosphorus stratification is that biological activity can now increase an already high phosphorus solubility situation due to the excessive concentration of phosphorus from surface applications. This increase in phosphorus solubility will become a problem if rain events exceed the soil infiltration rates and water runs off the field into the waterways.

Another issue is the nightcrawler channels. I have demonstrations showing how effective nightcrawlers can help drain your soil. The demonstrations are done by digging up a tile and putting a smoke bomb in the tile and covering the tile back up. After a few short minutes, smoke begins to filter up through earthworm channels all over the area affected by the tile. If phosphorus solubility is increased because of concentration and microbial activity, it seems to me that the earthworm channels are pipelines to waterways. This phosphorus is leading to excessive algae blooms and toxic algae in waterways and lakes.

The fix is one that neither the state agencies nor the farmers really want, and that is why it is probably a good fix. If phosphorus will not move in the soil, then farmers need to mechanically move it lower in the soil profile. Banding and deep-placing the phosphorus 5-8 inches deep would require some very expensive equipment. The best, easiest, and cheapest method would be to moldboard plow the fields every 5 or 6 years—maybe even longer if the research proves it. This suggestion makes many people go practically crazy. I have come from the time when plowing was done every year and the soil erosion was horrendous. I am not talking about continuous plowing. I believe if 10-15 percent of the fields were plowed following a wheat crop, the algae problem could be fixed in 5-6 years. By plowing after a wheat crop in late summer, the soil could be worked down and planted to a cover crop to minimize the erosion problems. The plowing would do two things: the phosphorus would be rolled deeper into the soil, getting it off the surface, and it would shear the earthworm channels temporarily. By moving the phosphorus deeper in the soil, the following crops would be less affected by dry surface conditions. The earthworms would re-establish their channels, but the high concentration of phosphorus would be off the surface, where rain events would be less of a problem.

Phosphorus Sources

Type	% P$_2$O$_5$	%P	#/ton P$_2$O$_5$
Mono Ammonium Phosphate	52	23.0	1040
Di Ammonium Phosphate	46	20.0	920
10-34-0 Liquid Polyphosphate	34	14.8	680
Rock Phosphate	25	11.0	500
Bone Meal	22	9.7	440
Fish Meal	8.2	3.6	164
Crab Meal	3.6	1.6	72
Chicken Manure w/o Litter 45% DM	2.4	1.1	48
Chicken Manure w/ litter 75% DM	2.2	1.0	45
Turkey Manure w/o Litter 29% DM	0.8	0.35	16
Swine Manure no bedding 18% DM	0.5	0.2	10
Beef Manure on concrete 15% DM	0.4	0.2	8
Compost Depending on the source	0.5-6.9	0.2-3	10-138

Table 10

Building phosphorus levels in the soil starts first with whether you want to go organic or commercial (inorganic). There are pluses and minuses to either way. The commercial approach will use the top 3 phosphorus sources in the above list and possibly rock phosphate, if available and priced right. Many commercial farms are now buying chicken and turkey manure to replace the typical phosphorus sources. Beef, dairy and swine manure contain too much liquid to truck very far away. Farmers buying manure are generally paying about half the cost per pound of commercial phosphorus fertilizer. Organic farmers will be restricted to any of the resources below the top three in the list. The above chart will help you decide how much phosphorus fertilizer will be needed to meet your goals.

Correcting Phosphorus Issues

At the beginning of this section I listed some ideal levels to start with. Assuming your soil report is below these suggested levels, it is relatively easy to figure out how much you will need. Even though it may be easy to figure out what you need, paying for it may be a whole different story. I'm not going to get into finances, but one of my mentors once reminded me that the soil I wanted to balance grew a crop last year and will more than likely grow a crop this year. That soil didn't get out of balance overnight and it will take time to bring it back.

All it takes to improve low-fertility soils is to apply more nutrients than you remove. While doing this, do your best to improve organic matter and biological activity. Start by choosing a realistic goal. Remember that excess nitrogen will be detrimental to organic matter. Looking at Table 4 or other crop removal charts will set the minimum level of nutrients that you need, and then adding some extra will set you on the path to balancing your soil.

Going the organic path provides some unique challenges to balancing your soil. As you have seen in the charts, compost, manures, or other organic products do not just contain one nutrient such as phosphorus. The organic sources contain all the major elements and more. The availability can also be more restrictive. It is a lot more difficult to balance your soil using organic products since the high concentration of nutrients are just not there compared to the commercial products. The tendency is to end up with a lot more phosphorus in the soil. Growing a good corn crop requires a huge amount of nitrogen, but using manure as the primary nitrogen supplier provides much more phosphorus than crop removal. This is great while rebuilding an abused piece of soil, but when you get to the ideal soil levels, excess phosphorus begins to build. Organic farmers are using a lot more tillage to control weeds, so soil loss is a concern with or without excessive P levels.

Rock phosphate and bone meals are some of the highest organic sources of phosphorus available, but being able to get them in large quantities and at an affordable price may be a bit of a problem. Both are low-soluble products and will work much better when the soil pH is below 7—preferably below 6.5. Given a choice between Tennessee brown rock phosphate and Florida rock phosphate, I prefer the Florida rock phosphate since it is very finely ground. It is much better to have low-soluble products finely ground; this will increase the surface area and increase the availability. Powdered products create a spreading problem and require a drop spreader or incorporation into a compost, or something of similar consistency, so as to reduce the dust and provide a larger volume for spreading with manure or lime-type spreaders. The Tennessee brown rock is coarser and probably could be blended with potassium sulfate or other granular-type materials. The rock phosphates will carry calcium and a variety of trace elements. I have used rock phosphates at rates of 1000 lbs./ac. for a one-time application, but smaller application rates could be used. It is important to do a very good job of incorporation. I would not recommend using rock phosphates in no-till situations.

Once you have reached your desired phosphorus level in the soil, use those products containing a blend of NPK to replace your phosphorus removal rates, and if you still need more nitrogen, switch to the purer nitrogen sources. For the organic grower, using sources like blood meal and feather meal would be preferable and the commercial grower could use any of the pure forms in the market.

I don't see any problem with applying enough phosphorus for two-year crop removal since phosphorus is very stable in the soil. I would do this only after getting to the desired phosphorus levels.

Chapter 6

Calcium

- **Ideal soil level:** Low-exchange soils (<10) 1200-2500 lbs./ac. at a 6-inch depth
- **High-exchange soils:** (>10) 65 percent base saturation, 70 percent base saturation on soils with clay content above 60 percent.
- **Mobile in the plant:** No
- **Xylem mobile:** Yes
- **Site of initial deficiency symptoms:** New leaves
- **Role in the plant:** Critical for root development, cell wall integrity, nitrogen metabolism, seed germination, pollination, fruit set.
- **Deficiency symptoms:** Calcium is not mobile in the plant, once incorporated in the plant structure it cannot be remobilized. Therefore, the deficiency symptoms are limited to the new growth. Leaves will be small and show possible necrosis on the leaf tip and edges. New leaves will curl down and develop a claw-like appearance. Tillering in cereal grains may be reduced and the stems soft and limp.
- **Toxicity symptoms:** Calcium toxicity is relatively unheard of. Calcium will raise the pH and affect other nutrients through reduced solubility or direct interference on the uptake of other cations.

General Discussion

Calcium is a vitally important nutrient for the growth of plants. I can't say it is the most important nutrient since all the essential nutrients are important, but calcium ranks right up there with nitrogen when it comes to driving the ship of production. Calcium not only has dramatic influence on the structure of the plant, from its roots to the plant itself; it can also have a positive influence on improving soil structure, especially on the higher-exchange-capacity soils. No other nutrient has as much effect on plants and the soils it grows in.

Before discussing the soil side of calcium, let's look at its effect on plant development. Calcium is first and foremost the backbone of cell wall integrity. Roots and foliage need a constant supply of calcium from seed germination to fruit set. Without calcium, seeds will germinate, but that first root development phase will not happen. Calcium must be at the growing point of plant roots. Behind the root cap is a zone of cell proliferation, and calcium must be pumped into each cell before fluid is injected into the cell for the elongation of cells. Without the calcium for strong cell walls, elongation or the movement of roots through the soil would be very minimal. At the first bit of resistance, the cell wall would collapse, stopping root expansion. Since calcium is only xylem mobile, it travels in the water transport system and goes only one direction—up. Calcium cannot be foliar-fed to help root development. For that matter, calcium left on the top of the soil through a lime application cannot help root development outside the zone of calcium concentration. Even with adequate calcium levels in the soil, high applications of potassium and nitrogen as ammonia (NH_4) can interfere with the uptake of calcium. Manure applications can exacerbate a calcium deficiency, as can commercial fertilizers.

Calcium is essential for the movement of metabolites in and out of the cell. The translocation of carbohydrates into the seeds or fruits requires calcium. Shortages of calcium during the growing season will affect the cell wall of, say, an apple, and determine how well that apple fills with carbohydrates. A shortage of calcium during the cell wall formation will lead to soft skin and easy bruising during handling. Foliar feeding of calcium can help in these situations. One of the easiest signs of calcium deficiency is blossom end rot. This can happen on any of the soft tissue fruits and vegetables like tomatoes and peppers. After a blossom is pollinated there is a limited time in which calcium is pumped into the fruit, and the bottom cells of the fruit is the last place calcium is deposited.

Any limitation of calcium results in the bottom of the fruit having weak cell walls, and as the fruit expands during development, these weak cell walls become thinner and thinner, subject to diseases and rots. The bottom of the fruit is subject to staying wetter and is likely to have soil splashed from rain or irrigation accumulate on the bottom, which contain bacteria and fungi, exacerbating the disease issue.

Blossom end rot can happen even on fairly good calcium levels for a couple of reasons. If the plant is flowering and still growing, the calcium sink is not the new fruit being formed, but the new vegetative growth. Calcium is traveling in the xylem and is being pulled by the evapotranspiration of the leaves, bypassing the fruit to a degree. Blossom end rot is more than likely to occur on the first couple fruit clusters and then go away. Foliar feeding calcium at the first sign of flower will help prevent this issue. Another situation that could result in calcium being limited to the new fruits and foliage is high humidity conditions from excessive rain. Since evapotranspiration drives the movement of xylem-mobile nutrients, calcium and boron would be severely impacted by a long rainy spell.

High levels of nitrogen in the soil tend to increase growth rate and increase the demand for calcium. If enough calcium is not available, the plants will experience soft growth and will be subject to lodging and diseases. Insect pressure from sucking insects like aphids might occur.

Calcium levels in the tissue for vegetables should not drop below 0.8 percent. Below this level, the new growth will exhibit brown veins, small leaves with burnt leaf tips, and the claw-like appearance of the new leaves. Legumes with calcium deficiencies will show chlorotic blotches on the newer growth.

Cereal grains will tiller less under low calcium conditions and roll the leaves with burnt tips much the same as you would see with copper deficiency. The stalks will be soft and subject to lodging. Sometimes calcium deficiency will show up as crinkled leaves.

Field beans exhibit flower abortions under calcium shortages, substantially reducing the yield potential.

Adequate or even high levels of calcium on the standard soil test does not guarantee that plants will receive enough calcium. We will cover this issue in more depth in the paste section. Below are the most common sources of calcium in use today.

Calcium Sources

Type	% Ca	Relative Neutralizing Power #
Calcitic Lime	30-35	85-100
Dolomite Lime	20-25	95-110
Gypsum	23	0
Burnt Lime	55-65	150-180
Rock Phosphate	33	
Bone Meal	22-26	
Calcium Nitrate	20	
Wood Ash	25-45	
Poultry Layer Manure	3-4	

Table 11

Percent calcium in Table 11 is based on pure calcium carbonate. Virtually all the above sources could be considered organic, but these should be verified by the organic certifiers.

Correcting calcium issues on low-exchange soils (<10)

Soils with exchange capacities of less than ten should be balanced using the SLAN approach. I would start with 1200-2500 pounds per acre depending on the exchange capacity. Let's look at the following example and see what might be the best approach.

These are very low exchange capacity soils with both calcium and magnesium deficiencies. The desired levels for this report is set with calcium being 68 percent base saturation and the magnesium with a sufficiency level of 200 pounds per acre. These low-exchange-capacity soils should be balanced, covering the magnesium levels first. Remember that these low-exchange soils do not have much clay that can negatively be

Sample Location			House	Pasture
Sample ID			Garden	1
Lab Number			98	99
Sample Depth in inches			6	6
Total Exchange Capacity (M.E.)			4.31	2.50
pH of Soil Sample			5.4	5.3
Organic Matter, Percent			1.86	2.64
ANIONS	Sulfur	p.p.m.	9	13
	Mehlich III Phosporous	as (P_2O_5) lbs / acre	66	27
EXCHANGEABLE CATIONS	CALCIUM lbs / acre	Desired Value	1173	680
		Value Found	719	326
		Deficit	-454	-354
	MAGNESIUM lbs / acre	Desired Value	200	200
		Value Found	131	73
		Deficit	-69	-127
	POTASSIUM lbs / acre	Desired Value	200	200
		Value Found	156	176
		Deficit	-44	-24
	SODIUM	lbs / acre	29	40
BASE SATURATION %	Calcium (60 to 70%)		41.68	32.59
	Magnesium (10 to 20%)		12.66	12.16
	Potassium (2 to 5%)		4.64	9.02
	Sodium (.5 to 3%)		1.48	3.47
	Other Bases (Variable)		6.60	6.80
	Exchangeable Hydrogen (10 to 15%)		33.00	36.00
TRACE ELEMENTS	Boron (p.p.m.)		0.27	0.27
	Iron (p.p.m.)		159	176
	Manganese (p.p.m.)		19	40
	Copper (p.p.m.)		0.77	0.53
	Zinc (p.p.m.)		1.49	2.99
	Aluminum (p.p.m.)		609	903
OTHER				

Table 12: *Soil Report*

impacted by higher magnesium levels. On the garden soil, we are 69 pounds short of magnesium and with a pH at 5.4; it is pretty much a no-brainer that we should use dolomite lime first. Dividing the magnesium deficit of 69 pounds by 11 percent (magnesium concentration in dolomite lime), we need 627 pounds of dolomite lime to meet the magnesium deficit. These 627 pounds will also supply approximately 132 pounds of calcium. The remaining calcium deficit is 322 pounds of calcium (454-132). This means that nearly 1,073 pounds of high-calcium lime (322 divided by 30 percent, the concentration of calcium in high-calcium lime) would be needed to finish supplying the remaining calcium deficit. If a choice had to be made to spread one lime over the other, I would start with the dolomite and finish up with the high-calcium lime at a later date. Incorporating the lime to at least 6 inches will speed the balancing project and will help root development.

Earlier I said it was a no-brainer to use dolomite lime to supply the magnesium, but what if dolomite is not available? Magnesium sulfate and KMag could be used in place of the dolomite lime. The concentration of magnesium in these two products could be divided into the magnesium deficit, as was done with the lime to determine the amount of product needed. High-calcium lime could be used to supply the calcium and adjust the pH. This will be reviewed further in the magnesium section.

The plant availability will all depending upon the coarseness of the lime, moisture level in the soil, and yearly rainfall amounts. In parts of the country that receive large amounts of rain during the spring and growing season, much of the available calcium could be lost and could result in a shortage of soluble calcium for plant growth. Soils that tend to be on the dry side may not be moist enough to dissolve the needed calcium for plant growth. The paste analysis would indicate whether these conditions might exist.

Correcting calcium issues on high-exchange-capacity soils (>10)

I recommend balancing soils with an exchange capacity of greater than ten using the Albrecht method of Basic Cation Saturation Ratios. Why use this method over the SLAN approach? The SLAN approach is only interested in meeting a specific number of pounds per acre per cation, and it is not interested whether there may be an excess level of cations—just shortages. These high-exchange soils are high as a result of having a large amount of organic matter or clay. These two parameters are the only fractions within soils that contribute to the exchange capacity. Excess amounts of magnesium, potassium, and sodium can lead to the breakdown of soil structure, resulting in soils that are tighter and stickier. This results in soils having low infiltration and percolation rates. When this occurs, anaerobic conditions will persist, damaging the good biological activity in the soil. Soil nitrogen will also be lost due to denitrification. Below is a picture of the soil texture triangle and where the SLAN and BCSR work best. The soils falling in the area below the red line, where there are high levels of silt and sand, are where the BCSR will not work and the SLAN approach is the method of choice for balancing soils. The soils between the red and yellow line are those with exchange capacities between 9 and 10, where a combination of the two balancing techniques should be applied. Be sure to balance the cations, but make sure the sufficiency levels are also met.

I have talked to many people who thought their soils were heavy but only had exchange capacities of 8 or 9. These soils were high in silt. Both silt and sand virtually have no charge and contribute very little to the overall exchange capacity. Silt particles fall somewhere in size between a fine sand and clay. Since they have no charge, it is not possible to flocculate the particles of silt as with clay. These soils will not drain particularly well and are difficult to work with. The best you can do for high-silt soils is to maximize organic matter by adding plant residues and compost. Growing cover crops with aggressive root systems like cereal rye and rye grasses really helps.

The area above the red and yellow line is soils with enough clay in which the BCSR will work well. It is in these soils that balancing the soil cations to 65-70 percent calcium, 10-15 percent magnesium and 3-5 percent potassium will maximize the flocculation of clays. This balance in the soil chemistry will allow for better overall soil structure and optimum soil biology. Of course, growing cover crops and adding organic matter is always a good thing, no matter the soil texture.

Soils high in organic matter—8-10 percent or more—will show less benefits from the BCSR approach to balancing soils because the organic matter is improving the soil structure by itself. When in doubt, I would balance the cations whenever the exchange capacity is greater than ten.

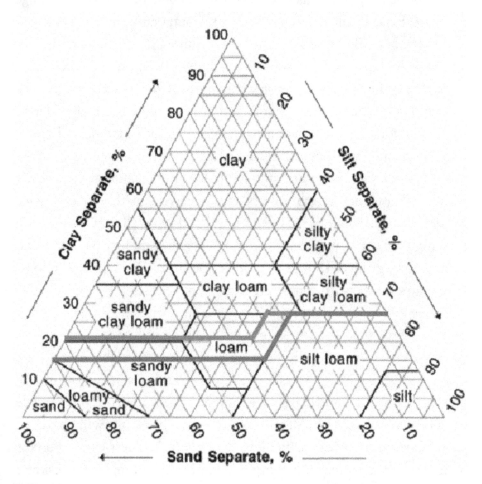

Table 13

It is easy to fix a calcium deficiency when the pH is low. First, look at the calcium and magnesium percent base saturation. If the magnesium is below 15 percent, then a portion, or all, of the lime will be dolomite lime. Always start your cation balancing with magnesium since it can have a detrimental effect if it gets too high. See the sample below. Looking at the S111 sample, the magnesium deficit is 16 ppm. Since the sample depth is six inches, we divide the depth by 3 inches (every 3 inches of soil equals one million pounds); multiply times 16 gives us 32 pounds of magnesium deficit. Logan sets their desired levels for magnesium at 12 percent, but I prefer using 15 percent as my desired level. I could just add in a little more dolomite lime in my recommendation or recalculate the desired level of calcium and magnesium myself. The exchange capacity is 11, so by multiplying 11 by 15 percent I get 1.65 meq. of magnesium. One meq. of magnesium is equal to 240 pounds at 6 inches; therefore, 1.65 meq. of magnesium is 396pounds per acre, instead of the 159 ppm or 318 pounds on the test.

Pounds per acre of each cation per milli-equivalent (Meq.) at various depths.

Cation	3 inches	6 inches	9 inches
Calcium	200	400	600
Magnesium	120	240	360
Potassium	390	780	1170
Sodium	230	460	690
Hydrogen	10	20	30

Table 14

Now the deficit is 396 – 284 (value found ppm x 2), or 112 pounds per acre. Dividing the 112-pound deficit by the concentration of magnesium in dolomite lime (11 percent), I need 1,018 pounds of dolomite lime to correct the deficit. Depending upon the fineness of the lime and level of incorporation, this deficit may correct quite rapidly, or it may take a year or two. Now I can look and see how much high-calcium lime I will need to add to correct the calcium deficit. Logan sets the base saturation of calcium at 68 percent, which is fine, but I prefer 65 percent, so let's reset the calcium desired value. Not resetting the desired value might push the pH up a little higher than I want. Multiplying the 11-exchange capacity by 65 percent means that I will need 7.15 milli-equivalents of calcium to obtain a 65 percent base saturation. One meq. of calcium is equal to 400 pounds; therefore 7.15 times 400 pounds means that my desired value will be 2,860 pounds instead of the 1,502 ppm or 3,004 pounds of calcium. The 1,018 pounds of dolomite lime will supply 214 pounds of calcium (1,018 x 21 percent calcium in dolomite lime). 2,860 - 214 equals 2,646 pounds of calcium that I still need; this can be made up with an additional application of high-calcium lime. If our source of high-calcium lime is 30 percent calcium, then 2,646 pounds divided by 30 percent means that an additional 8,820 pounds of calcium will be needed to satisfy the calcium deficit. The two limes would total nearly 9,838 pounds of lime. This is would be too much lime to add in one combined application. Adding this amount of lime could affect phosphorus, potassium, and trace element availability. I would add the 1,018 pounds of dolomite lime and maybe 2,500-3,500 pounds of calcium lime for one season and apply the rest of the calcium the following year. The level of incorporation would/should help determine the size of the lime application. I do not consider moldboard plowing or chiseling to be a thorough method of incorporation. I would not add all the calcium lime one year and then the dolomite the following year since it would really spread the calcium and magnesium ratio too wide, possibly inducing some magnesium deficiencies.

The A-1 sample on Table 14 is a low pH sample with high magnesium. It might appear as though an application of high-calcium lime is all that is needed. In this case, if we met the calcium deficit with an application of high-calcium lime, the pH would go above our desired level—and we haven't even begun to correct the potassium deficiency. Remember that the soil pH is a product of all the cations, not just calcium and magnesium. Our goal on A1 is to eventually bring the magnesium base saturation down to 15 percent and at the same time raise the calcium to 65 percent. We can start this process by adding enough high-calcium

Calcium

Sample Location			S111
Sample ID			
Lab Number			64
Sample Depth in inches			6
Total Exchange Capacity (M.E.)			11.04
pH of Soil Sample			5.4
Organic Matter (percent)			3.40
ANIONS	SULFUR:	p.p.m.	16
	Mehlich III Phosphorous: ppm		441
EXCHANGEABLE CATIONS	CALCIUM: ppm	Desired Value	1502
		Value Found	995
		Deficit	-507
	MAGNESIUM: ppm	Desired Value	159
		Value Found	142
		Deficit	-16
	POTASSIUM: ppm	Desired Value	172
		Value Found	121
		Deficit	-50
	SODIUM:	ppm	46
BASE SATURATION %	Calcium (60 to 70%)		45.04
	Magnesium (10 to 20%)		10.75
	Potassium (2 to 5%)		2.82
	Sodium (0.5 to 3%)		1.80
	Other Bases (Variable)		6.60
	Exchangeable Hydrogen (10 to 15%)		33.00
TRACE ELEMENTS	Boron (p.p.m.)		0.43
	Iron (p.p.m.)		289
	Manganese (p.p.m.)		13
	Copper (p.p.m.)		9.24
	Zinc (p.p.m.)		89.39
	Aluminum (p.p.m.)		932
OTHER			

Table 15: *Soil Report*

Sample Location			A1
Sample ID			
Lab Number			5
Sample Depth in inches			6
Total Exchange Capacity (M.E.)			12.52
pH of Soil Sample			5.8
Organic Matter, Percent			7.37
ANIONS	SULFUR:	p.p.m.	6
	Mehlich III Phosphorous:	as (P_2O_5) lbs / acre	161
EXCHANGEABLE CATIONS	CALCIUM: lbs / acre	Desired Value	3405
		Value Found	2460
		Deficit	-945
	MAGNESIUM: lbs / acre	Desired Value	360
		Value Found	651
		Deficit	
	POTASSIUM: lbs / acre	Desired Value	390
		Value Found	184
		Deficit	-206
	SODIUM:	lbs / acre	30
BASE SATURATION %	Calcium (60 to 70%)		49.12
	Magnesium (10 to 20%)		21.67
	Potassium (2 to 5%)		1.88
	Sodium (0.5 to 3%)		0.52
	Other Bases (Variable)		5.80
	Exchangeable Hydrogen (10 to 15%)		21.00
TRACE ELEMENTS	Boron (p.p.m.)		0.43
	Iron (p.p.m.)		354
	Manganese (p.p.m.)		37
	Copper (p.p.m.)		2.97
	Zinc (p.p.m.)		2.43
	Aluminum (p.p.m.)		570
OTHER	Ammonium (p.p.m.)		2.7
	Nitrate (p.p.m.)		4.5

Table 16: *Soil Report*

lime to raise the pH to just over 6.2. I would start by adding high-calcium lime to fill around 75 percent of the deficit, or around 600 pounds of calcium. Dividing 600 by 30 percent would mean making a high-calcium lime application of 2,000 pounds per acre. Keeping your liming applications between one and two tons per acre, even on high-exchange soils, is a good idea. Ultimately we want to adjust the pH, but supplying enough soluble calcium is the priority.

The balance of calcium needed should be supply through the use of gypsum. The sulfur is low, so this would also help to remedy this issue. I prefer to limit the gypsum applications to 1,000-1,200 pounds per acre, primarily due to the cost factor. By adding the lime and the gypsum, calcium will replace some magnesium on the colloid and will start to rebalance the base saturation. The high-calcium lime alone will not supply an anion that results in a highly soluble compound as gypsum does. Magnesium carbonate is not nearly as soluble or mobile as magnesium sulfate. The sulfate from the gypsum aids in the movement of the magnesium out of the soil profile. Cations or anions do not move out of the soil profile in water without maintaining an electrical balance. It is critical for the soils to have internal drainage in order to move out the excess magnesium; otherwise the magnesium may precipitate out on the compacted layers, making internal drainage poorer. Deep tillage may be required to break up these layers, providing an avenue for the excess magnesium to move out.

Balancing calcium and magnesium is relatively easy when the pH is low, but when the pH is high due to excess levels of cations it is a little more difficult. The above soil test is a prime example of this issue. It is more common for me to see high pH levels due to excess situations than to see low pH levels due to deficiencies. The above P3 sample has a high pH due to both magnesium and potassium. In a situation such as this, there is common practice in the field of soil balancing to always use high-calcium lime until the base saturation of calcium is above 60 percent, regardless of the pH level. In today's farming practices, I don't think this is practical or wise. Thirty years ago we did some of these things, but to make them work there was a lot of heavy tillage like plowing, offset disking, and deep tilling being practiced. We might apply

Sample Location			P 3
Sample ID			
Lab Number			57
Sample Depth in inches			6
Total Exchange Capacity (M.E.)			14.01
pH of Soil Sample			7.4
Organic Matter (percent)			5.76
ANIONS	SULFUR: ppm		22
	Mehlich III Phosphorous: ppm		257
EXCHANGEABLE CATIONS	CALCIUM: ppm	Desired Value	1905
		Value Found	1481
		Deficit	-424
	MAGNESIUM: ppm	Desired Value	201
		Value Found	340
		Deficit	
	POTASSIUM: ppm	Desired Value	218
		Value Found	1183
		Deficit	
	SODIUM: ppm		43
BASE SATURATION %	Calcium (60 to 70%)		52.83
	Magnesium (10 to 20%)		20.19
	Potassium (2 to 5%)		21.65
	Sodium (0.5 to 3%)		1.32
	Other Bases (Variable)		4.00
	Exchangeable Hydrogen (10 to 15%)		0.00
TRACE ELEMENTS	Boron (p.p.m.)		0.75
	Iron (p.p.m.)		137
	Manganese (p.p.m.)		33
	Copper (p.p.m.)		2.32
	Zinc (p.p.m.)		9.67
	Aluminum (p.p.m.)		317
OTHER	Cobalt (p.p.m.)		0.189
	Molybdenum (p.p.m.)		0.03
	Selenium (p.p.m.)		0.2
	Silicon (p.p.m.)		23.4
	EC (mmhos/cm)		0.37

Table 17: *Soil Report*

3 tons of high-calcium lime prior to going to corn and use 500-600 pounds of ammonium sulfate as part of our nitrogen source. We then would follow this practice up a year later with 1,500 pounds of gypsum. It worked. I had a father and son client who had a field that was giving them fits when it came to getting good production because the field was so tight and sticky. The pH was 7.5 with calcium around 50 percent and magnesium over 35 percent. When I sat down with the father and son to go over the soil reports, I told them that I wanted to do just what I described above and wanted to start out with 3 tons of high calcium. The father leaned over the soil report and pointed out that the pH level was 7.5. I told him that I knew that, but I wanted to try and mass flow some of the excess magnesium out of the soil profile and readjust the calcium and magnesium balance. Apparently I was not very convincing, because he pushed back from the table, got up, and told me this was a bunch of crap—but not quite in those words. The son indicated that Dad was a bit old fashioned and they would do whatever I recommended since this field was such a problem. About five years later I was driving past the field, and the father was taking wheat off. I reluctantly stopped and he stopped the combine and motioned me over. I really thought I was going to get chewed out, but the father said "Look at this." He dug his heel into the ground and pointed his finger at me and said that five years ago it would have taken an axe to get in this ground. Five years ago I was full of it, and now a guy who knew nothing about soil chemistry and soil balance was elated with the way the ground was preforming. This experience truly sold me on the idea of soil balance.

However, this approach to balancing soils today is not practical because of the lack of tillage that farmers are willing to do and the unwillingness to spread upwards of 500 pounds of ammonium sulfate. This approach to balancing high pH soils in minimum tillage situations would definitely have a negative effect on nutrient availability, especially since we have used glyphosate, a chelating agent, for nearly thirty years and the rainwater is no longer acidic.

Applying high-calcium lime to soils today with a high pH due to high magnesium and calcium levels below 60 percent may be causing more problems than it's fixing. Lime applied to high-pH soils, regardless of the fineness, will drop in solubility to around 10 percent or less. This means that there will be raw lime or free carbonates sitting on the surface of the soil in a reduced tillage situation. It will not be on the soil colloid. If you put lime on a soil as described above and then resample a year later, the odds are very good that the calcium levels on the soil test will be above the magic number of 60 percent; but what really happened? The industry standard for extracting solutions in soil testing is called Mehlich 3. This extracting solution has a pH of 2.5. It is more than likely that the Mehlich extracting solution will dissolve the free carbonates sitting in the top of the soil profile and give you a higher calcium number on the test. This might make you feel better, but did it really impact the soil? The impact on the soil structure will only occur when calcium has replaced magnesium on the colloids and flocculation of the clays has taken place. If there are enough free carbonates in the soil sample, the Mehlich extracting solution could get neutralized and reduce its effectiveness for extracting the other nutrients.

My approach to correcting the calcium issue on sample P3 is 1,000 pounds of gypsum for two or three years and adding the more acidic nitrogen sources when needed. I would also like to do some subsoiling in order to provide a pathway for excess magnesium and potassium to move out of the root zone. Cover cropping with a deep-rooted cereal rye would also be helpful.

Table 18 is a perfect example of the extracting solution dissolving extremely large amounts of calcium found in various calcareous soils located around the country. These soils have a high pH level and cannot be balanced by either the SLAN or BCSR approach. The only way to deal with this situation is to use the paste test to see what is really in solution. Some people are extracting these soils using an ammonium acetate extraction solution with a pH level of 8.2. All the cations are analyzed using the 8.2 extracting solution, but all other nutrients are analyzed using the Mehlich 3 solution. The 8.2 extracting solution will not dissolve free carbonates in the soil, but it tends to underestimate the potassium and magnesium. The paste analysis will show us basically what the plants see in solution, so I think that is about as good as we are going to get. I have been very successful using this method on these types of soils.

Chapter 7

Magnesium

- **Ideal soil level:** Low Exchange soils (<10) 220-400 lbs./ac. at 6-inch depth. Magnesium base saturations could be as high as 20 percent.
- **High-exchange soils (>10):** 10-15 percent base saturation; the higher the clay content of the soil, the lower the magnesium percentage.
- **Mobile in the plant:** Yes
- **Xylem and phloem mobile:** Yes
- **Site of initial deficiency symptoms:** Older leaves, but not necessarily the very bottom leaves.
- **Role in the plant:** Chlorophyll formation, phosphorus movement in the plant, the maturity of fruit; involved in over 300 enzyme functions in plants; promotes the germination of pollen.
- **Deficiency symptoms:** The visual symptoms almost always involve the chlorosis of the older leaves and then move to the newer growth. For grasses, the deficiency symptom appears as light striping. Dicots generally have intercostal chlorosis and blotchiness of the older leaves. Reddish to brown margins will occur on some plants like cauliflower and cabbage. Pines will show yellow tips on the lower part of the tree and proceed to the base of the needle.
- **Toxicity symptoms:** Excess magnesium like calcium doesn't cause toxicity per se, but it can interfere with the uptake of other cations and manganese. Damage to the soil structure on high-exchange soils with a sufficient amount of clay is possible.

General Discussion

Magnesium is the central atom at the heart of the chlorophyll molecule and cannot be replaced in that capacity by any other ion. Without magnesium, so goes the circle of life. Only about 15-20 percent of all the magnesium is tied up with chlorophyll. The remaining magnesium is involved in a multitude of enzyme systems; not the least of these is aiding in the ADP to ATP photosynthetic energy transfer in the plant. Magnesium also helps the uptake and movement of phosphorus within plants.

Generally, legumes need about twice as much magnesium as grasses. Plants that produce large amounts of starch, like potatoes and sugar beets, also have a high demand for magnesium.

Low levels of magnesium on cereal crops will have a negative effect on the test weight of grain. Early magnesium deficiencies during vegetative growth may not have a large impact on yield if the deficiency is corrected early enough. Take wheat, for example. If the deficiency is corrected before the second to last leaf and the flag leaf emerges, yields will not be significantly impacted since the last two leaves provide almost all of the carbohydrates for the seed head.

Magnesium, along with calcium and potassium, affects the plasma levels in the cells regulating the water budget of the plant. Deficiency of any of these cations will cause the plant to enter into drought stress much more rapidly. This is especially critical on low-organic-matter soils, which hold less water to start with.

With the exception of the magnesium in the chlorophyll molecule, magnesium can to some degree be replaced by calcium, manganese, cobalt, and zinc in the other enzyme systems, but not to the efficiency level of magnesium. The magnesium requirement of plants is not as great when manganese is in sufficient levels.

Nitrate nitrogen will stimulate the uptake of magnesium while ammonia, calcium, and potassium can reduce the uptake. Calcitic lime applications on acidic soils can temporarily interfere with magnesium uptake. Using smaller amounts of lime and spreading out repeated applications can help reduce interference issues. This is very important on the low-exchange soils.

Calcium to magnesium ratios on low-exchange soils might go as low as 3:1, but on the higher-exchange soils; somewhere between 4.5 and 6:1 seems to work the best.

The addition of sulfate ions from ammonium sulfate, gypsum, or other sources will enhance the leaching of magnesium from the soil.

Magnesium sources

Type	% Mg
Magnesium Sulfate (Epsom Salts)	17
Dolomite Lime	11
KMag (Potassium Magnesium Sulfate)	11
Magnesium Oxide	54
Magnesium Oxysulfate	36
Wood ash	3.5

Table 18

Correcting magnesium on low-exchange soils (>10)

The SLAN approach should be used to correct magnesium deficiencies on low-exchange soils. Due to a limited buffering capacity on low-exchange soils, remember that whatever nutrient we use to correct a deficiency can very easily affect some other nutrient in a negative way. Table 11 shows two low-exchange soils with magnesium deficits. These can be corrected a couple of different ways. For example, the house garden could be corrected with a dolomite lime, which would probably be the cheapest. Dividing the house garden's 69 pound magnesium deficit by 11 percent (the concentration of magnesium in the lime), we would need 627 pounds of dolomite lime to correct the magnesium deficiency. Additional high-calcium lime would be needed to finish off the calcium shortage. What if dolomite lime were not available? A high-calcium lime and KMag, also known by its old name of sul-po-mag, could be used to help fix the magnesium deficiency as well as the potassium deficiency. The deficit of 69 pounds should be divided by 11 percent (the concentration of magnesium in KMag), resulting in the need to apply 627 pounds per acre to correct the deficit. The KMag would also supply the needed sulfur and nearly 140 pounds of much-needed potassium for a vegetable garden. It would take nearly 400 pounds of magnesium sulfate, which would also correct the sulfur, but a potassium source would need to be found.

Correcting magnesium deficiency on high-exchange soils

The method for correcting a magnesium deficiency on a high-exchange-capacity soil consists of using the Basic Cation Saturation Ratio. The goal is to supply enough magnesium to bring the cation saturation ratio to 15 percent. I often find soils with magnesium in excess of 15 percent, but that will be covered a little bit later in this section. Table 21 shows three soil samples: two with magnesium deficiencies and one just about as good as it gets. These soils have been on soil balancing programs for over thirty years, and this is the first time that dolomite lime was recommended. Up to this time, only high-calcium lime had been used. This report does not have any desired values, so they have to be calculated. Starting with sample one, the exchange capacity is 10.75 and we want 15 percent of the soil colloids to be saturated with magnesium. 10.75 times 15 percent equals a desired value of 1.6 meq. of magnesium.

Sample Location			House	Pasture
Sample ID			Garden	1
Lab Number			98	99
Sample Depth in inches			6	6
Total Exchange Capacity (M.E.)			4.31	2.50
pH of Soil Sample			5.4	5.3
Organic Matter (percent)			1.86	2.64
ANIONS	SULFUR:	p.p.m.	9	13
	Mehlich III Phosphorous:	as (P_2O_5) lbs / acre	66	27
EXCHANGEABLE CATIONS	CALCIUM: lbs / acre	Desired Value	1173	680
		Value Found	719	326
		Deficit	-454	-354
	MAGNESIUM: lbs / acre	Desired Value	200	200
		Value Found	131	73
		Deficit	-69	-127
	POTASSIUM: lbs / acre	Desired Value	200	200
		Value Found	156	176
		Deficit	-44	-24
	SODIUM:	lbs / acre	29	40
BASE SATURATION %	Calcium (60 to 70%)		41.68	32.59
	Magnesium (10 to 20%)		12.66	12.16
	Potassium (2 to 5%)		4.64	9.02
	Sodium (0.5 to 3%)		1.48	3.47
	Other Bases (Variable)		6.60	6.80
	Exchangeable Hydrogen (10 to 15%)		33.00	36.00
TRACE ELEMENTS	Boron (p.p.m.)		0.27	0.27
	Iron (p.p.m.)		159	176
	Manganese (p.p.m.)		19	40
	Copper (p.p.m.)		0.77	0.53
	Zinc (p.p.m.)		1.49	2.99
	Aluminum (p.p.m.)		609	903
OTHER				

Table 19: *Soil Report*

Pounds per acre of each cation per milli-equivalent (Meq.) at various depths.

Cation	3 inches	6 inches	9 inches
Calcium	200	400	600
Magnesium	120	240	360
Potassium	390	780	1170
Sodium	230	460	690
Hydrogen	10	20	30

Table 20

The soil depth on the report shows 9 inches. Therefore, 1.6 Meq. times the pounds per acre of magnesium at 9 inches equals 576 pounds (1.6 x 360). The desired level for magnesium is 576, and subtracting the value of 350 leaves a deficit of 226 of magnesium. With a low pH, the magnesium source of choice would be dolomite lime. Dividing 226 by the magnesium content of dolomite lime (11 percent) means that 2,054 pounds of lime would be needed to correct the magnesium deficiency. The calcium in the dolomite lime would give us about 300 pounds of extra calcium, but that's assuming all the lime would dissolve. Apply the lime, incorporate, and retest in a year.

Sample 2 in Table 21 has an exchange capacity of less than 10, so should we use the SLAN or the BCSR approach? This is one of those samples on the borderline where both should be considered. At first glance, the calcium appears to exceed the desired SLAN top end level and the magnesium is in-between the desired level, but this sample was collected at 9 inches—not 6 like the pre-set desired levels. Since this sample was collected deeper, the upper limits for calcium, magnesium, and potassium should be increased by nearly 30 percent. This would mean the calcium SLAN level should be 3,250 pounds, the magnesium at 520, and the potassium at 550-650, depending upon the crop. Since the pH is below 6.5 and the exchange capacity is below 10 and could go a little higher when I neutralize the hydrogen ion content, I would still go ahead and balance the soil based on the SLAN approach. The desired

			1	2	3
Sample Location			1	2	3
Sample ID					
Lab Number			17	18	19
Sample Depth in inches			9	9	9
Total Exchange Capacity (M.E.)			10.75	8.45	12.32
pH of Soil Sample			5.9	5.7	6.4
Organic Matter (percent)			2.54	2.68	2.89
ANIONS	SULFUR:	p.p.m.	13	12	27
	Mehlich III Phosphorous:	as (P_2O_5) lbs / acre	236	546	234
EXCHANGEABLE CATIONS	CALCIUM: lbs / acre	Value Found	4062	2817	4983
	MAGNESIUM: lbs / acre	Value Found	350	279	653
	POTASSIUM: lbs / acre	Value Found	416	420	419
	SODIUM:	lbs / acre	77	59	84
BASE SATURATION %	Calcium (60 to 70%)		63.01	55.56	67.40
	Magnesium (10 to 20%)		9.05	9.17	14.72
	Potassium (2 to 5%)		3.31	4.25	2.91
	Sodium (0.5 to 3%)		1.04	1.02	0.99
	Other Bases (Variable)		5.60	6.00	5.00
	Exchangeable Hydrogen (10 to 15%)		18.00	24.00	9.00
TRACE ELEMENTS	Boron (p.p.m.)		0.37	0.26	0.48
	Iron (p.p.m.)		124	119	125
	Manganese (p.p.m.)		45	16	44
	Copper (p.p.m.)		1.6	2.31	1.93
	Zinc (p.p.m.)		4.18	3.42	3.1
	Aluminum (p.p.m.)		592	656	560
OTHER					

Table 21: *Soil Report*

levels for calcium and magnesium, just based on 65 percent calcium and 15 percent magnesium, would be 3,296 and 456 pounds respectively. This is still very close to the SLAN-desired levels. It would be best to use paste analysis to determine if we need to push the numbers to the upper limits. My goal is to meet the levels in the paste and still keep the pH in the 6.5 range. It is best to go slow on the low-exchange-capacity soils since it is easy to overdue the lime and get the pH too high. The other concern should be how well this will lime be incorporated and to what depth. If it is only going to be incorporated 6 inches and we are balancing for 9 inches, it would be best to put on maybe half the lime this year and the other half next year. Balancing the magnesium would require nearly 1,609 pounds of dolomite lime to correct the 177-pound magnesium deficit. This application of dolomite lime would also supply 378 pounds of calcium, which would be roughly 100 pounds of calcium short. The remaining 100 pounds of calcium could be supplied by approximately 400 pounds of gypsum. The exchange capacity could raise almost 2 milli-equivalents from the summation of the additional cations after correcting the calcium and the magnesium.

Table 22 is the test of a soil comprised of free carbonates. Free carbonates mean that there is some lime in the soil that is not dissolved. In this case, it means that there is a very high amount of raw lime contained in the soil. This soil comes from Texas and it is a natural phenomenon found in a large portion on their soils. In Florida, the corral-based soils will look very much the same way. Any soil that has ever been limed could have a certain amount of raw lime in the soil profile. This could be due to coarseness of the applied lime or a pH level, which is high, reducing the solubility of the lime. This soil is rather light and sandy, but the exchange capacity might lead you to believe this is a very heavy clay. The Mehlich extracting solution, which has a pH level of 2.5, is dissolving this lime, adding large amounts of calcium to the extracting solution. Upon analyzing the concentration of cations in the solution with the ICP unit in the lab, the cations are summed to calculate the total exchange capacity. The result is an overestimated exchange capacity and outlandish desired values in these types of soils. Applications of 300 to 400 pounds of KMag in these instances has given tremendous

Sample Location			North
Sample ID			
Lab Number			1
Sample Depth in inches			6
Total Exchange Capacity (M.E.)			54.95
pH of Soil Sample			7.6
Organic Matter (percent)			2.44
ANIONS	SULFUR:	p.p.m.	20
	Mehlich III Phosphorous:	as (P_2O_5) lbs / acre	53
EXCHANGEABLE CATIONS	CALCIUM: lbs / acre	Desired Value	14945
		Value Found	18231
		Deficit	
	MAGNESIUM: lbs / acre	Desired Value	1582
		Value Found	1429
		Deficit	-153
	POTASSIUM: lbs / acre	Desired Value	1714
		Value Found	894
		Deficit	-820
	SODIUM:	lbs / acre	83
BASE SATURATION %	Calcium (60 to 70%)		82.95
	Magnesium (10 to 20%)		10.84
	Potassium (2 to 5%)		2.09
	Sodium (0.5 to 3%)		0.33
	Other Bases (Variable)		3.80
	Exchangeable Hydrogen (10 to 15%)		0.00
TRACE ELEMENTS	Boron (p.p.m.)		0.71
	Iron (p.p.m.)		45
	Manganese (p.p.m.)		153
	Copper (p.p.m.)		2.49
	Zinc (p.p.m.)		7.52
	Aluminum (p.p.m.)		213
OTHER			

Table 22: *Soil Report*

results in plant response. Soils this far out of balance would be nearly impossible—and certainly cost prohibitive—to attempt to balance out right. The best thing to do is to try and balance the soil solution. When we discuss paste analysis, we will continue on with this sample.

Magnesium removal by crop

Crop	Pounds for average crop yield
Cabbage	10
Cauliflower	3
Celery	27
Lettuce	5
Onions	14
Tomato	35
	Pounds per bushel
Corn	0.09
Beans	0.21
Wheat	0.15

Table 23

Chapter 8

Potassium

- **Ideal soil level:** Low-exchange soils (<10): 220-400 lbs. of K/ac at 6-inches. 6-10 percent K base saturation. 220-300 pounds for pastures, 300-400 pounds general crops; 400-500 pounds specialty crops, alfalfa, vegetables.
- **High-exchange soils (>10):** 400-650 pounds of K/ac. 4-6 percent K base saturation; 250-300 pounds for pastures; 350-450 pounds general crops; 450-600 pounds specialty crops alfalfa, vegetables.
- **Mobile in the plant:** Yes
- **Xylem and phloem mobile:** Yes
- **Site of initial deficiency symptoms:** Older leaves.
- **Role in the plant:** Carbohydrate metabolism along with mobilization; regulates water utilization through the stomata on the leaves, acts as a summer coolant and winter antifreeze, affects color and sweetness in fruit, helps in disease resistance through the metabolism of sugars to carbohydrates.
- **Deficiency symptoms:** Potassium does so much in the plant that a deficiency may have many faces. Deficient plants will slow in growth and will eventually become stunted and have shorter internodes. Since potassium is so mobile in the plant, a deficiency will tend to show up on the older leaves. Grasses exhibiting a potassium deficiency will show up as burning on the outside edges of the leaves. Other plants will exhibit burning at the leaf tip on the older leaves, with the base of the leaf remaining green. Plants with a K deficiency will often times wilt more rapidly in the heat of the day. Plants will tend to lose their lower leaves as the deficiency progresses, leaving crops like alfalfa stemmy and low in protein. Diseases and insect pressure may increase since sugars may accumulate in the tissue of the leaves or roots. Potassium can act as an antifreeze in the plant, helping them to resist

freezing with large temperature swings in the winter.

Grains of potassium deficient plants will be smaller and lighter in test weight. Vegetables like tomatoes may be small and off color. The shoulders of the tomatoes will often times be yellow especially on the later setting fruit. Greens may be bitter. Bruising is more prevalent on low potassium fruit especially potatoes, tomatoes, peppers and tree fruit.

- **Toxicity symptoms:** Excess potassium will primarily result in driving soil pH levels high and causing a salt burn on the tissue. Excessive potassium applications can lead to uptake interferences of calcium, magnesium and ammonium ions as well as the trace elements of boron, manganese and zinc.

General Discussion

Soil compaction will limit root mass and affect the uptake of all nutrients. Potassium is one cation that seems to be exceptionally affected since it can be fixed or trapped between clay layers. Soils with excessive magnesium, which are tight and sticky, exacerbate the fixing of potassium. Once soils are brought back into balance, often times extra potassium starts showing up on the soil test. Over liming will limit potassium uptake; excessive potassium will limit the uptake of calcium and magnesium cations. Magnesium and potassium can easily interfere with the uptake of each other. Potassium to magnesium ratio should be around 2:1 and as high as 3:1 in the soil. This ratio could be as low as 1.5:1 on modified soil mixes for potted plants, raised beds and hoop houses. It is the ratio of 2:1 potassium to magnesium in KMag that makes it such an ideal product of choice for balanced soils and pastures. Excessive nitrogen in the form of ammonium(NH_4) can reduce potassium uptake. This issue will exacerbate plant diseases along with root rots. Therefore, a good rule of thumb is, as you increase population and nitrogen rates, potassium should also be increased. Nitrogen to potassium ratios in the plants should run close to 1:1 for cereal grains and specialty crops (vegetables) and as high as 1.5:1 for other crops like corn or cannabis and as high as 2:1 for oil seeds like soybeans.

Potassium in young plants need to be on the higher side, due to the high metabolic demand of the expanding new tissue. Older tissue will accumulate potassium as long as there is an adequate supply. Deficiencies in the tissue will almost always give depressed yields if it occurs prior to flowering.

Some clays have a lattice structure in which potassium is just the right size and fits into these openings. When potassium slips into these openings, another clay particle can be pulled down over the opening, fixing the potassium and making it unavailable to the plants. It may look as though your potassium application was lost, but it's just bound up in the clay structure. In these types of soils it is best to make your potassium applications as close to planting as possible. Potassium will continue to be fixed until the lattice openings are filled. Your soil report may not improve for some time until many of the openings are filled. Keeping the calcium base saturation between 65 and 70 percent will help limit the potassium fixation. Dry soils will also have a detrimental effect on potassium uptake.

Potassium not only helps to regulate turgor pressure and water in the plant—it also contributes to the salt concentration in the plant, acting as an antifreeze and helping plants winter over or get through periods of frost. Foliar feeding potassium prior to a freeze event will help reduce frost damage to the plants.

High organic soils or modified soil mixes do not hold potassium very well. Potassium is quick to move in solution, so high applications will overwhelm the soil solutions, blocking calcium and magnesium. We will look at this situation in the saturated paste section.

Potassium is key in the metabolic process of converting sugars into starches. Excess sugars in the plants will attract insects, especially the sucking kind like aphids and leaf-hoppers. The intrusion into the cell also opens up the plants to diseases. Of course, fungal and bacterial diseases are quite aggressive in their own right and don't necessarily have to have an entry point from an insect. A secondary problem of aphids is the production of honey dew that they deposit on the plant leaves. When this material molds, it turns black, reducing the photosynthetic capabilities of the leaves. We learned in the sulfur section that sulfur helped reduce excess nitrogen in the plants. Deficiencies of both sulfur and potassium will exacerbate a disease and insect problem. Excessive nitrogen applications will only add to this problem.

Plants that use a high amount of nitrogen for production generally need at least that amount of potassium, even though the crop removal may be relatively small. For example, 180 bushel of field corn may need 240 pounds of nitrogen and potassium for production of the corn crop, but only 50 pounds of potassium will be removed by the corn. Conversely, a 40-ton tomato crop needs roughly 230 pounds of nitrogen and double that amount of potassium to produce the crop. Nearly 290 pounds of potassium will be removed by the crop. Generally, vegetables are potassium hogs and need a constant supply in their fertilization program. Trying to supply the potassium needs strictly from manure applications will more than likely create an excess phosphorus situation.

Just a quick note about potassium and turf grass for lawns: once the potassium is optimized in the soil for our lawn, virtually no more potassium or phosphorus needs to be added as long as the grass clippings are mulched back on the lawn. Removing the clippings will remove a lot of potassium, some phosphorus and calcium, and a lot of nitrogen. So, if you are not into doing a lot of yard work other than mowing the grass, keep the clippings on the yard.

Potassium Sources

Type		% K$_2$O	%K	lbs./ton K$_2$O
Potassium Chloride	0-0-60	60	50	1200
Potassium Carbonate Liquid	0-0-30	30	24.9	600
Potassium Acetate Liquid	0-0-20	20	16.6	400
Potassium Sulfate	0-0-50	50	42	1000
Potassium Nitrate	14-0-44	44	36	880
KMag Potassium Magnesium Sulfate		22	19	440
Kelp Meal		10	8	200
Alfalfa Meal		2.4	2	48
Chicken Manure w/o Litter	45% DM	1.7	1.4	34
Chicken Manure w/ litter	75% DM	1.7	1.4	34
Turkey Manure w/o Litter	29% DM	0.7	0.6	14
Swine Manure no bedding	18% DM	0.2	0.17	4
Beef Manure on concrete	15% DM	0.5	0.4	10
Horse Manure	46% DM	0.7	0.6	14
Green Sand (Glauconite)		5	4	100
Compost Depending on the source		1-3.6	1-3	20-72
Wood Ash		3.6-8.4	3-7	72-168

Table 24

Organic farmers have a good choice of high-concentrated soluble potassium fertilizer sources—unlike with nitrogen and phosphorus. With the exception of the first three on this list, all could be potential organic sources. Greensand is a commonly used potassium source, but it is a very low-soluble product and should be considered as a long-term source.

Some people make the claim that soils have all the potassium that you would ever need—10-20 thousand pounds per acre. This is true, but almost all of it is locked up in the clay structure of the soil.

Correcting potassium issues on low-exchange soils

Correcting potassium deficiencies will be done much the same as you would phosphorus deficiencies. You apply more than the crop removal. There is only one catch. Applying excess phosphorus, even in low-exchange soils, will tie up with a whole host of nutrients and remain in the soil. That is not the case for potassium or the other cations like calcium and magnesium. Potassium and the other cations will attach to the negative colloid sites of the clays or humus. It is a continuous competition between the various cations and the exchange sites. It is desirable to try and achieve a balance between these cations where it is possible, mainly on high (>10) exchange capacity soils. Using the Basic Cation Saturation Ratio (BCSR) will not work on low-exchange-capacity soils. Therefore, the Sufficiency Level of Available Nutrients (SLAN) is the approach that must be used on low-exchange soils. Our job is to apply enough nutrients, especially the cations, in quantities and balance to satisfy the growing crop.

On really low-exchange-capacity soils, most of the crop requirements for potassium will have to be completely supplied. Looking at the crop removal chart for 25 ton of potatoes, the crop would need 264 pounds of K_2O for the tubers and another 90 pounds of K_2O for the vines, making a grand total of 354 pounds of K_2O. This is of course assuming that a 25-ton potato yield is realistic. A 190-bushel corn crop would need only 51 pounds of K_2O for the grain, but 209 for the stalks and leaves. Before applying the pure forms of potassium, I would consider what sources that I might use for nitrogen and phosphorus to see if there is any potassium attached to these products. Organic farmers might use manures and compost as their nitrogen and phosphorus sources, all of which have potassium attached to the products. Commercial growers would commonly stick to pure NPK sources unless the availability of manure is close and priced right. Finding a manure source for low-exchange soils would be beneficial because of the slower release of nutrients and the organic matter benefits. A potato crop needing roughly 350 pounds of K_2O would amount to nearly 700 pounds/ac. of potassium sulfate and 600 pounds/ac. of potassium chloride. Seventy five percent of this potassium would leave with the potato crop. There is no way a low-exchange soil could hold an application such as this, besides the other nutrients needed for the crop. This large of a potassium application could also block calcium and magnesium uptake. Split applications prior to and during the growing season is one way to minimize the interference issues, but in doing this increases labor and management dramatically. Applying dry fertilizer to a growing crop is difficult at best without burning the foliage, and injecting potassium liquids requires very large rates.

On the other hand, a 190-pound corn crop would need 260 pounds of K_2O per acre, of which nearly 80 percent of the potassium would remain on the farm with the stalks. This would be equivalent to nearly 400 pounds/ac. of potassium sulfate or 350 pounds/ac. of potassium chloride left on the field for the following crop. Applying the 600 pounds/ac. of potassium chloride for a potato crop on low-exchange soils could quite possibly result in some chloride burn. Residue management as well as the use of cover crops is especially critical on low-exchange soils. Therefore, growing a crop like potatoes, which have a high affinity for potassium, might be best preceded by a corn or a cover crop that leaves a large amount of potassium behind in the residue. Potassium in the plant does not become part of the structure like calcium, so it will wash out of the tissue before the residue totally decays. This practice can also apply to high-exchange soils. Generally, this type of nutrient management for a following crop requires tillage to maximize the nutrient release for a high-demand crop.

Correcting potassium issues on high (>10) exchange soils

On high-exchange soils, we can have nearly enough available potassium on the soil colloids for the high-affinity crops like alfalfa, beets, potatoes, and corn. It would be ideal to elevate the base saturation of potassium to around 5-6 percent for these demanding crops. A 15-exchange capacity soil would have nearly 700 pounds of exchangeable potassium, but all the exchangeable potassium might not be available to the crop. High-exchange soils, which are high due to clay content, are prone to more compaction issues. This would limit root mass and potassium availability. It is also very unlikely that you could reduce the potassium on the colloids to much less than 1.5 percent. Plants take up nearly all their potassium from the soil solution. Once the potassium gets so low on the colloid, the solution would become so diluted that the plant roots would fail to function properly. Dry weather would also hinder potassium uptake, especially if much of your potassium tended to be stratified in the upper portions of the soil profile. Soils with a good base saturation of potassium should still have a portion of the crop removal amount applied before the growing season, until you are comfortable knowing that enough potassium will be available for the crop during the season. Tissue analysis would help you make this judgment. I like to see all high-exchange soils have a base saturation of 4 percent—even growing crops that have a low affinity for potassium. Pastures with a high degree of grasses need to have the soils balanced properly with calcium and magnesium, otherwise the grasses will luxury-feed on the potassium, possibly resulting in grass tetany issues. Timothy is one grass that doesn't seem to luxury-consume potassium like the fescues and ryegrasses.

When trying to rebuild soil potassium levels, don't try to make the fix all in one year. Potassium will interfere with the uptake of other cations, especially magnesium. If you need to add a large amount because of the next crop in rotation, you might want to add a little gypsum and KMag to temporarily keep the soil solution balanced until equilibrium has been reached in the soil.

Chapter 9

Sodium

- **Ideal soil level:** >50 lbs./ac. for unresponsive crops, < 100 lbs./ac. for crops such as wheat, oats, beets, celery, carrots, spinach, and cotton that are responsive to sodium in the soil. This works best when the potassium levels are adequate. Keep the sodium base saturation less than potassium.
- **Mobile in the plant:** Yes.
- **Xylem and phloem mobile:** Yes
- **Site of initial deficiency symptoms:** In low-response plants, there is none. High-response plants like sugar beets and mangolds will demonstrate a thinning of the leaves. The leave may have a metallic green coloration with a purple tinge on the underside of the leaves.
- **Role in the plant:** Sodium can work in helping maintain turgor pressure in the plant, similar to potassium. In some plant species, sodium will replace potassium to a certain degree. In beets and celery, sodium can have a positive effect on root development and yields.
- **Toxicity:** Toxicity in plants is generally related to salt damage. Things like leaf margin and tip burn will show up along with a squatty appearance in growth. Low-sodium tolerant plants (natrophobic) such as corn, cucumbers, sunflowers, lettuce, onions, and strawberries would do best if the sodium tissue levels were below 1500 ppm. Sodium-tolerant plants (natrophilic) such as beets may not show any issues until 40,000 ppm in the leaves. High levels in the soil tend to deflocculate clays, destroying the soil structure. This will result in poor drainage and aeration.

General discussion

Sodium is a nutrient that has been generally thought of as a problem child. It has certain benefits for some plant species, and others very little at all. Plants like corn, beans, cucumbers, lettuce, onions, and strawberries actually inhibit sodium uptake and translocation in the plant. This will cause the plants to expend energy and may result in the interference of other critical cations to be taken up. Plants like barley, oats, carrots, and tomatoes can take up fairly large amounts of sodium, but plants like beets, cotton, and spinach can take up extremely large amounts of sodium.

There is indication in the research that sodium applications will help those plants that tend to accumulate carbohydrates over those that accumulate proteins. This does not seem to apply to corn.

It has been suggested that crops that are responsive to sodium could benefit from an application of 75-100 pounds per acre of sodium in loamy soils and up to 150 pounds per acre in sands. As long as potassium is adequate, wheat and oats could benefit from an application of 50-75 pounds per acre. The high-sodium-uptake plants could benefit from applications of 75-150 pounds per acre. If sodium chloride was used as a sodium source, then chloride sensitivity would have to be investigated.

Sodium sources

Product	Concentration of sodium %
Sodium Nitrate	27
Sodium Chloride	40

Table 25

Sodium can replace potassium cations in the plant's metabolic processes in some plants when potassium is deficient. Plants like barley, alfalfa, clover, carrots, tomatoes, and asparagus could benefit from a sodium application in this situation.

Chapter 10
Chlorine

- **Ideal soil level:** Chlorine is typically not done on a standard soil test and guidelines are very limited. Paste test or solution data suggests keeping chloride levels between 20 and 60 ppm depending on the plant species. Some species (see Table 26) could tolerate higher levels, but 20-30 ppm in the paste should supply the nutritional needs of plants.
- **Mobile in the plant:** Yes
- **Xylem and phloem mobile:** Yes
- **Site of initial deficiency symptoms:** Wilting at the margins of younger leaves is generally the first sign of chlorine deficiency. Severe deficiency will result in the curling of young leaves as well as leaves reduced in size. Leaves may appear blotchy and may have a bronze coloration on the underside of the leaves. Severe deficiency could also prevent fruit set.
- **Role in the plant:** Chlorine will affect the plasma content in the plants as well as the water efficiency of the plants. Chlorine is one of the anions that balances against the cation accumulation in the plants.
- **Toxicity:** Toxicity in plants is generally related to salt damage, especially along sidewalks, parking lots, and roadways where salt is used for de-icing. Toxicity is also common where poor-quality irrigation water is being used. It is especially prevalent in arid parts of the country and hoop houses where rainwater can't flush out the chlorides. Even fairly good irrigation water can lead to toxicity problems if the cycles are consistently short.

General discussion

The plant's need for chlorine is much greater than for other minor elements. Chlorine deficiency is rare due to the regular use of potassium chloride in commercial operations. Organic farmers should be more concerned, especially those in the heartland of the country, away from coastal and industrialize areas. Irrigation water will supply much of the chlorine needed by plants. Current rainwater analysis that I have done shows about a half a pound of chloride is added to the field for every acre-inch of water. Twenty inches of rainwater during the growing season would contribute nearly 10 pounds of chloride. This would be the chloride equivalent of adding 25 pounds of potassium chloride.

Chloride will impact nitrogen uptake, but before plants start showing a nitrogen deficiency a salt toxicity is more likely to occur. Chlorides can replace nitrates to a certain degree in plants, reducing nitrate build up. Chloride will accumulate in plants, with tissue analysis showing as high as 6-10 percent chloride in the plants.

Chlorides are more easily taken up by plants than other anions such as sulfate and phosphates. Chlorides can also be adsorbed from the air.

Too many chlorides in plants will reduce the sugar content in leaves and the starch content in potatoes.

There is a large difference in tolerance to chlorine among plant species (see Table 26).

Crop sensitivity to chlorides

Tolerant	Intolerant
Barley	Beans
Wheat	Onions
Beets	Potatoes
Corn	Tomatoes
Cauliflower	Peas
Brussel Sprout	Cucumbers
Radish	Melons
Lettuce	Tobacco
Spinach	Strawberries
Palms	Fruit Trees
	Berry Bushes

Table 26

It is best to use sulfate sources of nitrogen and potassium for those crops sensitive to chlorides. If potassium chloride is the only available source of potassium, consider making a fall application so the chlorides have a chance to leach out with the winter and spring rains before planting.

Seed germination may be impacted by chloride levels, even on those plants considered tolerant to chloride levels such as cauliflower, lettuce, and beets.

Chapter 11

Trace Elements

Boron

—

- **Ideal soil level:** 1-1.5 ppm
- **Mobile in the plant:** No
- **Xylem mobile:** Yes
- **Site of initial deficiency symptoms:** New growth.
- **Role in the plant:** Cell-wall formation, pollen grain germination, seed production, sugar translocation.
- **Toxicity:** The range between adequate levels and toxic levels can be quite small. The older leaves will show yellowing, necrosis, and blotches on the margins of the leaves that work their way in to the center of the leaves. Yields and plant size will be depressed. Toxicity on tomatoes will occur at the leaflet tips and curl inward. Flushing out toxic levels of boron will take 3-4 times more water than for flushing out sodium chloride.

General discussion

Boron is absorbed as an anion and moves like calcium in the water transport system of the plant in the xylem. Therefore, foliar feeding boron to a deficient plant is only good for the foliage and does nothing for the root system. High humidity conditions will limit leaf transpiration and hurt the uptake and movement of boron in the plant.

Boron is essential for many metabolic processes in the plant such as carbohydrate and nitrogen metabolism. The development of the cell walls in the plant depends dramatically on boron. In fact, nearly 50 percent of all the boron in the plant is found in the cell walls. This is very close to the same for calcium.

Boron is necessary for the sugar and starch formation in the fruits, due to its positive impact on the photosynthesis activity of the plant. Phosphorus is another critical factor in photosynthesis and when it is limiting, the demand for boron increases. In contrast, low boron will lead to low sugars and low vitamin B levels. Nitrates will accumulate under low boron conditions.

Boron can also help manage water utilization in the plant, depending upon the water availability in the soil.

Boron helps improve the uptake of calcium, phosphorus, magnesium, and potassium from the soil; however, high levels of potassium can exacerbate a boron deficiency. Calcium applications can also reduce boron uptake by increasing the pH and precipitating out boron in the soil as calcium borate. Increasing calcium in solution means an increase in boron is necessary as well.

Boron demand is high during flowering, and if it is limiting, flower abortions are likely. This can be a problem in soybeans or dicots in general since they need about ten times more boron in the plants compared to grasses or monocots.

Boron deficiencies will be more prevalent in sandier soils or soils with free calcium carbonates from natural occurrences or excessive liming.

Like calcium, boron deficiencies tend to show up in the new growth of plants as chlorosis and tip burn. Some plants will exhibit misshapen leaves and shortened internodes. The reduction of flowers and seed formation will result in lowering yields. Roots will also show an effect of boron deficiency by having restricted growth and a bristly appearance.

There are not many boron sources available to growers (see Table 27).

Boron Sources

Product	% Boron
Granubor Calcium Borate US Borax	14.3% OMRI
Solubor US Borax	20.5% OMRI
Borax	10%
Boric Acid	17%

Table 27

Boron deficiencies can be corrected by any of these products at 5-15 lbs./ac. It is best to broadcast boron due to the threat of toxicity, but I have used a quarter of a pound of actual boron in a row starter for corn at 2 inches over and 2 inches below the seed with no problem.

Boron tolerance varies drastically between plant species. Table 28 shows the degree of tolerance for some common plant species. For a more detailed listing, go to the US Borax website.

Boron Tolerance Table

Tolerant	Moderately Tolerant	Intolerant
Alfalfa	Potatoes	Fruit Trees
Cabbage	Tomatoes	Strawberries
Cauliflower	Peas	Raspberries
Carrots	Barley	
Lettuce	Oats	
Onions	Corn	
	Wheat	

Table 28

This table doesn't mean that boron should only be used on the tolerant species. Apple and pear trees with a boron deficiency will have misshapen fruit with external cork spots. Peaches will exhibit brown blotches and cork spots also. Raspberries will have feathery looking leaves and reduced bud. These low-tolerant species just mean that if the boron range in soils is 0.8 to 1.5, you might want to stay to the low side when balancing your soils. The paste analysis would be a better indicator since it will show what the boron level is in solution.

Iron

- **Ideal soil level:** 50-100 ppm.
- **Mobile in the plant:** No
- **Xylem mobile:** Yes
- **Site of initial deficiency symptoms:** New growth
- **Role in the plant:** Chlorophyll formation, oxygen carrier, cell division, and growth.
- **Toxicity:** Iron toxicity rarely occurs, except sometimes in rice production, where the saturated soil conditions can reduce iron, making it more available. Excess iron could result in phosphorus tie-up in low pH situations, but it is more likely to occur from over-fertilizing iron.

General Discussion

Iron must continually be absorbed into plants due to the poor mobility and the lack of translocation from old leaves to the new growth. It is for these reasons that the deficiency symptoms will always show up on the new growth. Nearly 50 percent of all the iron taken up in the plant will come from direct root intercept with the soil colloids. The remaining iron will be absorbed from the soil solution. Dicots such as soybeans can develop more root hairs and, along with exuding more hydrogen ions, increasing the plant's ability to pick up more iron in low-iron situations. Grasses or monocots do not have this unique ability.

Carbonates and bicarbonates in the soil will exacerbate an iron deficiency, but sulfate anions will not. Excessive applications of nitrogen and phosphates may induce iron deficiencies.

Normal iron levels in plants will range between 50 and 200 ppm. Extreme care must be taken when tissue sampling so as not to have any soil contamination on the plants, because even small amounts of dust on a plant or soil splashed onto the plant from rain will show up as high iron and aluminum in the tissue test.

Iron is involved in many enzyme systems, acting as a catalyst. A catalyst assists in speeding up chemical reactions but does not become part of the reaction. Many of the trace-element metals perform this function within plants.

Weather conditions such as extremely wet or dry can induce iron deficiencies. Saturated soils that are tight and compacted exacerbate iron deficiencies by preventing the reduction of iron within the root.

High organic soils can also chelate iron into complexes, making it unavailable to plants. Plants that exhibit iron chlorosis rarely see the leaves die like deficiencies of other nutrients.

Fruit trees that produce heavily one year and not the next may experience iron deficiencies in the off year of production, because in the high-yielding year more assimilates move into the fruit and less into the roots. This reduces the root growth for the following year, causing less iron to be taken into the tree. Fruit tree order of decreasing susceptibility to iron is as follows: peach > pear > apple > apricot > cherry. Many trees don't initially show iron deficiencies until they are five years or older. The older the trees become, the further down the roots will move, into lower, unavailable iron levels in the soil. This is especially a problem with alkaline, glacial-till soils.

Due to the poor mobility of iron within plants, it is better to try and correct deficiencies through foliar feeding. In plants with extreme iron chlorosis, it is unlikely that you will be able to completely eliminate the deficiency symptoms. Iron chelates are generally a better choice for foliar feeding because they are less likely to be immobilized like iron sulfates would be. Using sulfates also increases the risk of a sulfate burn on the leaves. EDTA chelates will certainly be more expensive, but cheaper chelates like lignosulphonates and heptonates can be quite effective. The latter chelates will prove more difficult to keep in solution when blending multiple nutrients.

Iron Sources

Source	% Iron
Ferrous Sulfate	19
Ferrous Oxide	77
Chelated Iron	5-14
Iron Polyflavonoids	9-10
Iron Lignosulfonates	5-8
Azomite	1.4
Blood Meal	0.2-0.3

Table 29

Manganese

- **Ideal soil level:** 40-80 ppm, but extremely pH sensitive
- **Mobile in the plant:** Semi-mobile
- **Xylem and phloem mobile:** Yes
- **Site of initial deficiency symptoms:** The leaves midway up on the plant at the new growth. The deficiency symptoms will start higher up on the plant as compared to magnesium deficiency, which will start lower on the plant.
- **Role in the plant:** Involved in chlorophyll synthesis, many enzyme systems, improves calcium and phosphorus uptake, aids in disease resistance.
- **Toxicity:** Manganese toxicity is generally the result of low-pH soil environment — below 5.5. Levels in leaves above 300-550 are high enough to

result in toxicity. Some plants maybe higher or lower, but numbers in this range are a good indicator that problems may be just around the corner. Symptoms are brown specks on the leaves, browning of stalks, chlorotic lesions on the leaf tip, and crinkling of the leaves. Iron chlorosis could be a secondary issue due to manganese interference. Some plants may be more tolerant to excess levels of manganese in the soil due to their poor ability to absorb manganese. Plants that prefer to grow in higher calcium soils, such as alfalfa, will be more sensitive to toxicities.

Plant Tolerance to Excess Manganese

Tolerant	Intolerant
Potatoes	Beans
Rice	Cucumbers
Cotton	Lettuce
Oats	Cereals
Strawberries	Alfalfa
Timothy	Apple Trees
Clover	Cauliflower
	Tomatoes

Table 30

General Discussion

Manganese is absorbed in the reduced state (Mn^{+2}) and is much more prevalent in this state at lower pH levels, as well as in low oxygen situations. Conversely, over-liming and tillage will reduce manganese availability by oxidizing manganese to Mn^{+3}. Muck fields, which develop under saturated conditions, have very low levels of manganese due to the increase in solubility and loss of manganese over time during soil development.

I have seen carrots in the same field exhibit both deficiency on one end of the field and toxicity on the other end. The soil manganese level on both ends of the field was 10 ppm on the standard test. This is quite low by most people's standards, yet toxicity was visible on one end. The end where the toxicity existed had a pH of 4.5 which was deeper with the original muck. The opposite end is where the muck had played out and tillage had brought up marl and mixed it into the remaining muck. The pH on this end was 7.5 and deficiency symptoms were visible. Manganese availability increases about a hundred fold for every drop of one pH unit. The lesson learned was that pH can correct trace element deficiencies even to the point of toxicity, despite the levels in the soil.

Plant species will vary dramatically in their requirement for manganese. Most plants will grow normally when manganese levels in the tissues are greater than 25 ppm; however, vegetables will often need twice the normal tissue level. Manganese demand is greatest when the plants are rapidly expanding in vegetative growth and transitioning to reproductive growth; therefore, temporary deficiencies could occur during those times.

In an intense tissue sampling of soybeans being treated with glyphosate, I found that a glyphosate application essentially cut the level of manganese in the plant by half. At 40 ppm of manganese in the plant, the other plant nutrients started to decline. Manganese visual deficiency symptoms didn't occur until 25 ppm of manganese in the plant. The take-home message is that plants showing visual deficiency symptoms have more than likely been experiencing the deficiency effects long before the visual signs appeared. Soybeans receiving a glyphosate application should have manganese levels in the plant greater than 80 ppm. Two applications of glyphosate in the season should automatically have manganese added to the spray mix. I would assume that if glyphosate reduced manganese uptake that it would be possible to also reduce other nutrients as well, but I did not research that possibility.

Manganese mobility is low but it is much better than calcium and boron. Manganese cannot be mobilized from the lowest part of the plant like magnesium can, so the initial chlorosis will start midway up on the plants. Good levels of silicon and molybdenum will aid in the mobility of manganese.

Both manganese and magnesium are activators of many enzyme systems. Manganese is a key component in the oxidative/reduction reactions and is important in the chlorophyll and photosynthesis process. It is also beneficial in carbohydrate and sugar production in the plant. Plants low in manganese tend to be lower in vitamin C.

Plants with a manganese deficiency will always have a high iron content and vice versa. Manganese deficiency is almost always exhibited by chlorosis of the leaves for dicots and on grasses with a light green coloration and brown tips on the leaves. Grasses are less likely to have manganese deficiencies unless over-liming occurs. Take-all disease in cereals is dramatically increased from over-limed manganese-induced deficiency.

Manganese will influence the water utilization, like potassium, and will improve cold tolerance, similar to copper.

Loose, dry soils will enhance oxidation and exacerbate manganese deficiencies. Irrigation water high in bicarbonates will also limit manganese availability. Those conditions that roots tend to thrive on — loose, moist, well-aerated — soils tend to minimize manganese availability. Keeping the soil pH level in the 6.0-6.5 range definitely helps offset these negative availability factors. Excess levels of the divalent cations along with iron, zinc, and ammonia will compete with manganese uptake. A calcium-to-manganese ratio around 250:1 is considered normal for plants. Acid-forming nitrogen fertilizers and the chloride ions from potassium chloride will help increase the availability of manganese.

Looking at a plant from the top down and observing where trace element deficiencies would show up are as follows: iron would be on the newest under developed leaves, zinc would follow on the first newest mature leaf, and manganese would be lower but on the newer growth.

Products like kelp, Azomite, rock phosphate, etc. have some manganese in them, but in very small amounts and very low in solubility, especially for the rock dusts.

Manganese Sources

Product	% Manganese
Manganese Oxide	63
Manganese Carbonate	31
Manganese Sulfate	27
Chelated Manganese	10-12

Table 31

Copper

- **Ideal soil level:** 8-10 on mineral soils and probably higher on organic soils.
- **Mobile in the plant:** No
- **Xylem mobile:** Yes
- **Site of initial deficiency symptoms:** Structural parts of the plant, such as stocks and stems. Plants may appear soft and wimpy. Deformity in the seed heads.
- **Role in the plant:** Involved as a metabolic catalyst, important in photosynthesis as well as reproduction, helps improve flavor and sugar content, impacts nitrogen utilization as well as lignification.
- **Toxicity:** Copper toxicity will generally show up as chlorosis or reddish-brown necrotic lesions. Toxic levels of copper in the soil will block the uptake of other metals, giving the appearance of manganese, iron, or zinc deficiencies in the top growth.

General Discussion

Copper is absorbed as Cu^{+2} and, compared to boron, manganese, iron, and zinc, it is one of the lowest nutrients taken up by plants. However, copper deficiencies even below visual symptoms could lead to a reduction in yields by 20 percent. Copper in the soil is very insoluble and is tied to the soil colloids or bound up in organic complexes. It is for this reason that virtually all copper taken up by the plants is through direct root intercept with the colloids or organic matter. Any reducing root mass of plants, whether it be calcium deficiencies, root pruning from tillage or insects, compaction, or weather, will have a negative effect on the copper levels in the plant leaves. There could be plenty of copper in the roots but still deficiencies in the leaf tissue due to the poor mobility of copper within the plant. The copper status in plants should be determined by tissue analysis of the newer leaves.

Copper is important in cell wall integrity and lignification. It is during the early growth and development of the plant structure that the demand for copper is the highest. Deficiencies at this time will lead to weak and wimpy plants unable to withstand high wind conditions.

Copper is important in the nitrogen reductase enzyme system in the plant. Shortages will lead to the accumulation of nitrates, making the plants vulnerable to insects and diseases. It follows that as plants demand more nitrogen during growth, the need for copper also increases. Excessive nitrogen rates can exacerbate a copper deficiency. Plants such as alfalfa and beans, which need high levels of nitrogen for building proteins, require the highest levels of copper. Low protein levels in these plants could contribute to low copper. The appearance of yellowing and nitrogen deficiency could be caused by low copper.

Seeds harvested from copper-deficient plants do not germinate as well, especially under stress conditions. Most of these stress conditions involve cold growing conditions; copper has been shown to improve cold tolerance.

There is a close relationship of copper with nitrogen and it has been suggested that the Cu ppm / N percent be greater than 1.9.

One must be careful when correcting copper deficiencies with copper sulfate. Solutions over 0.2 percent copper sulfate may lead to leaf scorch. This is roughly 2.5 ounces in 10 gallons of water. Always test a foliar before spraying an entire crop. There is a wide range of sensitivities between various crops when spraying copper.

Copper toxicity in the soil is somewhat dependent upon the growing crop. Sugar beets are considered to be one of the most sensitive, with upper limits of 20-40 ppm, while barley can take 40-70 ppm, potatoes 50-70 ppm, lettuce 50-80 ppm, and wheat ranging between 80 and 100 ppm.

While visiting Portugal, I saw soil copper levels in their vineyards upwards of 300 ppm copper as a result of long-term usage of copper sulfate as a fungicide. Most of this copper would be on the surface due to the lack of tillage in the vineyards.

Hog manure is often high in copper, and repeated applications may result in elevated soil levels.

Copper Sources

Product	% Copper
Copper Sulfate	25
Copper Oxide	75-89
Copper Chelates	9-13

Table 32

Zinc

- **Ideal soil level:** 10-20, but very dependent on overall phosphorus levels.
- **Mobile in the plant:** No
- **Xylem mobile:** Yes
- **Site of initial deficiency symptoms:** Zinc will first be seen at the new growth of both root and shoot tips. The newest leaves will appear smaller than normal and yellow. Deficiencies will exhibit poor seed production. Overall plant will be short and squat.
- **Role in the plant:** Zinc will have a direct influence in chlorophyll production, auxin production, flowering and seed production, transformation of

sugars to starch and many enzyme systems such as RNA production and protein synthesis. It will help control cell division, especially at the root and shoot tips.
- **Toxicity:** Toxicity rarely occurs except around ore mine and slag discharge areas. High levels of zinc can be found near galvanized roofs or fences but it is rarely ever toxic. Liming and extra phosphorus applications will fix most high zinc situations.

General Discussion

In the soil, zinc is strongly bound by organic matter and clays. Certain clays like montmorillonite holds zinc much tighter than kaolinite clay. Zinc mobility in the soil is dramatically reduced by high organic matter, pH levels, and high phosphorus. This is more pronounced in sandy soils. High levels of soil phosphates will precipitate zinc out as a relatively insoluble zinc phosphate. Plants can absorb zinc phosphates, but phosphate accumulation in the roots will block the vertical movement of zinc to the leaves. High bicarbonate levels as a result of irrigation will greatly reduce the mobility of zinc within the plant. Cool soils early in the growing season, as well as dry soils latter in the growing season, will minimize zinc uptake and mobility.

Zinc deficiency in plants is primarily found on the new growth. A condition called "little leaf" is a result of zinc deficiency. The production of auxins is influenced greatly by zinc, and the lack of it produces very small leaves. Plant auxins are rapidly decomposed in sunlight, which is why the little leaf phenomenon is very pronounced in the citrus groves of the south. Zinc deficiency, as well as with other metals, is exacerbated by spraying glyphosate around the trees to control weeds.

Plants suffering from zinc deficiency are less cold tolerant and fail to produce enough lignum. This situation often leads to winter die back, whereas copper deficiency results in summer die back.

Plants deficient in zinc will tend to suffer from blossom drop, reducing seed production. Plants will vary in their degree of tolerance to zinc deficiency. Table 33 lists some of these plants, based on their ability to tolerate zinc deficiency.

Plant Tolerance to Zinc Deficiency

Low Tolerance	Medium Tolerance	High Tolerance
Corn	Potato	Rye
Beans	Beets	Peas
Grapes	Alfalfa	Carrots
Hops	Tomato	Asparagus
Oats	Onion	Pasture Grasses
Wheat	Spinach	
Barley	Lettuce	
Fruit Trees		
Citrus Trees		

Table 33

Manures can have high amounts of zinc, but they also have high levels of phosphates and organics, which tend to negate much of the gain. Swine rations tend to have high levels of zinc in the rations to prevent a skin disease called parakeratosis in pigs. High levels of zinc in the soil will impact the uptake of other metals such as iron, manganese, and copper.

Liming and phosphorus fertilization will reduce zinc availability. This can especially be bad in no-till situations, which will stratify the lime and phosphorus at the surface. Even though zinc may be reduced to the plant, it may not necessarily be reduced in the standard soil analysis, leading to incorrect interpretations. Tissue analysis would help to verify this assumption.

The addition of KMag and magnesium sulfate can help improve zinc availability by displacing it from soil colloids and organic sorption sites.

Sources of Zinc

Product	% Zinc
Zinc Sulfate	36
Zinc Oxide	78
Zinc Carbonate	52
Chelated Zinc	9-15
Zinc Lignosulfonates	5-10

Table 34

Molybdenum

- **Ideal soil level:** 1-3 ppm
- **Mobile in the plant:** Poorly
- **Xylem and Phloem mobile:** Yes
- **Site of initial deficiency symptoms:** New leaves will become narrow and deformed. Whip tail in brassicas is a good example. Chlorosis and necrosis of the newer growth.
- **Role in the plant:** Critical for the development of nitrate reductase system, which prevents the accumulation of nitrates. Molybdenum aids in the absorption of atmospheric nitrogen in the root nodules of legumes. It helps convert inorganic phosphorus to an organic form in the plant. Moly helps to increase amino acid levels in the plant.
- **Toxicity:** Toxicity is very rare but in studies where toxic levels were given to plants; a yellow to orange chlorosis appeared on the leaves. Most plants can tolerate relatively high levels. The toxicity comes when using plants high in molybdenum for animal feed. Levels in the plants greater than 5 ppm for sheep and greater than 10 ppm for cattle will cause diarrhea and interfere with copper and selenium utilization.

General Discussion

Compared to other trace elements, molybdenum is consumed in very small amounts. The removal rate of moly for 70 bushels of soybeans is roughly 0.1-0.3 ounces and for 210 bushels of corn is nearly twice the level of beans. A large tomato crop could remove upwards of 1.5 ounces. These numbers seem rather trivial compared to the other nutrients that we supply to growing crops; however, when an adequate soil level is 2-6 pounds, you can see that after a couple of decades of crop removal it is quite possible to start seeing molybdenum deficiencies.

Once the soil level for molybdenum drops below 0.7 ppm, deficiencies are likely to occur. This will be exacerbated by dry soils and soils with a pH levels below 7.2.

Moly is held quite well in the clay lattice structures and to aluminum and iron sorption sites.

Grasses are very capable of taking up molybdenum, making them less sensitive to deficiencies. This is quite possibly due to the fibrous root system. Dicots with taproots tend to be more sensitive to molybdenum deficiencies.

Sulfate and chloride anions will block moly uptake along with high levels of manganese, copper, and zinc. Phosphates and iron as iron sulfate can enhance the uptake of moly.

Molybdenum mobility in the plant is poor. That is why the moly levels within the plants are highest in the roots, then the stems, and finally in the leaves.

Deficiency of molybdenum in plants will visibly show up as chlorosis and/or necrosis of the younger and older leaves. Since moly is critical for nitrogen-fixing bacteria in legumes, a shortage could also show up as a nitrogen deficiency. Plants could test high in nitrogen, but that nitrogen could be the accumulation of nitrates, opening the plant up to disease and insect pressure. Plants short of moly will be lower in protein but higher in non-protein nitrogen (nitrates). This situation can be detrimental for both humans and animals. The more plants depend on nitrate nitrogen, the higher the demand for molybdenum.

Molybdenum not only effects amino acids and proteins in plants; it can also effect the carbohydrates and sugar levels in plants. The transfer of ADP to ATP, which is the energy in the plant, is influence by moly, and shortages will create an accumulation of sugars, causing the tissue to turn a red coloration, like a phosphorus deficiency does. Alfalfa, beans, spinach, cauliflower, beets, lettuce, tomatoes, celery, and citrus form a partial list of crops that respond well to molybdenum fertilization.

Sources of Molybdenum

Product	% Molybdenum
Ammonium Molybdate	54%
Sodium Molybdate (Dry)	39%
Sodium Molybdate (Liquid)	10%

Table 35

Cobalt

- **Ideal soil level:** 1-2 ppm
- **Mobile in the plant:** Poorly
- **Xylem and phloem mobile:** Yes
- **Site of initial deficiency symptoms:** A cobalt deficiency will look more like a nitrogen deficiency in legumes since cobalt is essential in the conversion of atmospheric nitrogen into nitrate nitrogen in the nodules.
- **Role in the plant:** Cobalt is essential for the development of vitamin B12 in the plant. It is a critical component of converting atmospheric nitrogen in the nodules to plant-available nitrogen.
- **Toxicity:** Toxicity is very rare, but in cases where excess cobalt has been added to the soil, iron deficiency has been induced, leading to yellowing of the new-growth leaves. Toxicity can be treated by raising the soil pH and increasing organic matter. Adding nickel will help reduce the symptoms.

General Discussion

Plants growing in high-pH, calcareous soils or over-limed sandy soils are likely to have low cobalt uptake.

Cobalt is taken up as a cation Co^{+2}. It is much like molybdenum in that the mobility is poor in the plant, leading to a higher accumulation in the roots, then the stems, and finally the leaves. In soils where cobalt was low, foliar feeding cobalt nitrate had a significant impact on the yields of beets, cabbage, and potatoes.

The deficiency of cobalt is more than likely to have a larger effect on humans or animals eating the plants than on the plant itself.

Cobalt sulfate is an inorganic source containing 31 percent cobalt.

Silicon

- **Ideal soil level:** 30-50 ppm
- **Mobile in the plant:** Poorly
- **Xylem mobile:** Yes
- **Site of initial deficiency symptoms:** Silicon deficiency is less visual than other nutrients. Deficiencies tend to be more in the area of disease and pest suppression as well as poor structural integrity.
- **Role in the plant:** Improve structural integrity of both the stems and leaves.
- **Toxicity:** No known toxicity issues.

General Discussion

In the past, silicon was considered a non-essential nutrient for plant growth, but new studies have shown it to be quite the opposite. The increased availability of silicic acid to plants have shown improved stem strength and disease resistance, especially for powdery mildew.

It has also shown suppression of sucking insects along with borers and hoppers.

Once taken up in the plant, silica is deposited in the epidermal cells of the plant, creating a barrier to invasion by diseases and plant pests. The deposited silica contributes to the structural soundness of plants like cereal grains.

Grasses and cereal grains are among the highest accumulators of silica, with rice being the highest.

Silicate ions have a negative effect of phosphorus uptake in hydroponic solutions, but it is quite the opposite in soils. Silica ions in the soils can displace phosphorus ions from absorption sites on colloids and organic complexes, forcing phosphorus to remain in solution.

Silicon availability in the soil is not affected by soil pH as are other nutrients. Plants such as melons, pumpkins, cucumbers, and wheat, which are susceptible to powdery mildew, could benefit from silica fertilization.

Even though sands are primarily made up of silica, they produce very low levels of available silica. Organic soils are also very low in available silica. Modified growing mediums used in greenhouses, hoop houses and pot production plants would all benefit from silicon fertilization.

Silica will help reduce the effects of over-fertilization of nitrogen fertilizer. It can improve both growth and yield. The barrier effect of silica at the cellular level will help plants during times of drought and cold weather.

Foliar feeding silicon is not very effective. Adding silicon to the soil or media mixes is by far the best way to get silica in the plant.

Since small grains are accumulators of silicon, the removal of the straw from the fields may remove as much as 40-50 pounds per acre of silicon.

Sources of Silicon

Product	% Silicon
Calcium Silicate $CaSiO_3$	24%
Magnesium Silicate $MgSiO_3$	28%
Potassium Silicate $K2SiO_3$	18%

Table 36

These silicon products all contain cations that when added to the soil will not only increase these cation levels, but the pH as well. Wollastonite is the name of a calcium silicate that has been used by organic producers. Crossover, which is a trade name by the Harsco company, is an interesting product. It is a calcium-magnesium silicate with 8.5 percent sulfur. Since many soils are deficient in sulfur, this might help solve a couple of problems. The calcium is listed as 25 percent, the magnesium as 3 percent, and soluble silica as 1.4 percent. In many of the dry silicate products, solubility may be very low, so most field applications may range from 1-3 tons per acre. There are liquid silicate products. Choose the highest concentration in order to get the biggest bang for the buck. Some powders are available for mixing in water, with varying degrees of solubility.

Heavy Metals

Often times I have been asked by people wanting to test their soil if they should test for heavy metals. My general response is no, unless you own or are buying a property in an industrial area or an old apple orchard. You will spend several hundred dollars for the appropriate testing.

The best analysis is through a certified environmental laboratory.

The analysis of choice is called the EPA 503 metals. There has been some work done on using the standard soil extracting solution, Mehlich 3, to indicate potential problems for lead, but if you want the best data possible I suggest getting the 503 metal analysis. Below is a list of the 503 metals and maximum allowable cumulative loading rates. These levels were set up for biosolids applications but provide acceptable ceiling levels for heavy metals per the EPA guidelines.

Parameters	Cumulative Loading Rates Kg/ha	Cumulative Loading Rates lb./ac.
Arsenic	41	36
Cadmium	39	34
Copper	1500	1300
Lead	300	260
Mercury	17	15
Molybdenum	N/A*	N/A*
Nickel	420	370
Selenium	100	89
Zinc	2800	2400

Table 37
*Not Available. *Ceiling concentrations for adding molybdenum containing materials is set at 75ppm, but no cumulative loading rates have been established.*

As you can see, the typical trace elements that we use in agriculture — copper and zinc — have very high loading rates. I have never seen the 503 metals for copper much over 120 lb./ac. or zinc over 480 lb./ac. on fields that have received biosolids since 1988. The Mehlich numbers on the standard soil test typically run around 5ppm and 15ppm for copper and zinc respectively. Most mineral soils with a minimal amount of clay have a great ability to fix metals. However, this example should be no reason to ignore potential problems in industrial sites or soils with a history of metal application.

The low limit metals, such as arsenic, cadmium, and mercury, are low because of their toxicity and potential for causing cancer. Lead, although significantly higher than the previous metals, is another metal to be taken seriously because of its adverse effect on human health.

Two things that will limit heavy metal availability is increasing organic matter and increasing the pH to over 6.5. Of course, limiting toxic heavy metal availability will also decrease the necessary trace elements for plant growth, such as zinc and copper.

Nickel, although listed in the heavy metal table, has recently been shown to have benefits for crop disease resistance and tolerance. As early as the mid 1940s, nickel was shown to benefit crops like wheat and potatoes in the area of disease protection. More recently, nickel deficiency has shown up in citrus as little leaf syndrome due to glyphosate tying up the metal. Excess copper and zinc can also interfere with nickel uptake. Pinching of the leaf tips, short internodes, and brittle stems can all be symptoms of nickel deficiency.

Selenium, although not considered essential for plant growth, is significant in improving animal health. Datnoff, Elmer, and Huber's book, *Mineral Nutrition and Plant Diseases*, is a great reference piece discussing all the major, minor, and the ultra-minor elements like nickel and their effect on plant health.

Chapter 12
Paste Analysis

IN SIMPLE TERMS, A PASTE ANALYSIS IS THE WATER EXTRACTION OF a soil or modified growing media. This process saturates the medium to a pancake batter consistency with distilled or irrigation water. The medium is held in this state for twenty-fou hours, at which time the water is vacuumed off and analyzed for nutrients, bicarbonates, chlorides and soluble salts. It is these parameters that the plants readily absorb. The bulk of nutrients are picked up through mass flow or diffusion. Copper and about half of the iron is picked up by direct root intercept (see Table 37).

Percentages of Nutrient Uptake Through Roots by Mass Flow, Diffusion, and Interceptive Root Growth

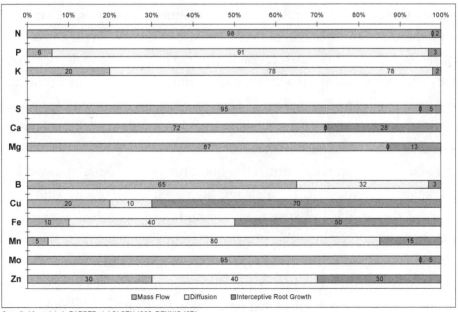

Compiled from data in BARBER abd OLSEN 1968, DENNIS 1971

Table 37

Since most nutrients are picked up out of solution, it makes perfect sense that paste analysis should be a standard part of our analytical soil test package. The paste analysis does a much better job of matching up with tissue analysis. In fact, the tissue analysis should be used to adjust the desired levels of the paste test. The guidelines below are just a starting point. Certainly, these numbers can be moved up or down according to tissue analysis and desired yields.

The paste data may or may not correlate to the data on a standard soil analysis. One might ask, "If tissue data correlates best to the paste analysis, why do both tests?" I feel that the tests complement each other and it is necessary to have both in order to balance the soil. When looking at a paste analysis alone, the one missing factor that we don't know is the flow rate of nutrients into solution. Even with a standard test we can't know the flow rate, but we can see what level each of the nutrients are at in the soil. It is somewhat like looking at a can and seeing how full it is. The higher the level in the can, the more nutrients available to the plant. We just don't know how big the hole is in the bottom of the can. If we have two cans, one just a third full and another clear full, and we drill a small hole in the bottom of each can, the can that is fuller will deliver its content faster due to the increased pressure from being clear full. The can with a lower amount could deliver the same amount of product over a given time, if the hole was bigger. The enigma is that we have no idea what the size of the hole is in our can. The size of the hole in the can or the movement of nutrients into solution is affected by the ratio of nutrients on soil colloids, the microbial activity, compaction, clay and organic content, and weather, just to name a few. We don't want a large amount of any one nutrient in solution; we want a balance, and that balance will be largely impacted by the forementioned factors.

Balancing the standard test is the first order of business, followed by balancing the paste test. If balancing high-exchange-capacity soils can improve things like soil structure, aeration, and microbial activity, it makes sense to prioritize this practice to improve nutrients in solution. The following two soil tests reflect the impact of flow rate into solution. These soils are from a sod farm and the rating of good or poor is based on the quality and volume of sod production in a split field. Looking at

			Poor	Good
Sample Location				
Sample ID				
Lab Number			33	34
Sample Depth in inches			6	6
Total Exchange Capacity (M.E.)			7.03	9.08
pH of Soil Sample			6.4	6.8
Organic Matter, Percent			2.10	2.91
ANIONS	Sulfur	p.p.m.	11	12
	Mehlich III Phosporous	as (P_2O_5) lbs / acre	269	591
EXCHANGEABLE CATIONS	CALCIUM lbs / acre	Desired Value	1913	2470
		Value Found	1844	2724
		Deficit	-69	
	MAGNESIUM lbs / acre	Desired Value	202	261
		Value Found	251	245
		Deficit		-16
	POTASSIUM lbs / acre	Desired Value	219	283
		Value Found	221	353
		Deficit		
	SODIUM	lbs / acre	51	50
BASE SATURATION %	Calcium (60 to 70%)		65.53	74.98
	Magnesium (10 to 20%)		14.87	11.24
	Potassium (2 to 5%)		4.03	4.98
	Sodium (.5 to 3%)		1.58	1.20
	Other Bases (Variable)		5.00	4.60
	Exchangeable Hydrogen (10 to 15%)		9.00	3.00
TRACE ELEMENTS	Boron (p.p.m.)		0.55	0.69
	Iron (p.p.m.)		185	208
	Manganese (p.p.m.)		82	87
	Copper (p.p.m.)		2.96	3.16
	Zinc (p.p.m.)		6.32	6.93
	Aluminum (p.p.m.)		461	623
OTHER	Ammonium (p.p.m.)		0.5	0.5
	Nitrate (p.p.m.)		1	1

Table 38: *Soil Report*

Sample Location			Poor	Good
Sample ID				
Lab Number			129537	129538
Water Used			DI	DI
pH			6.4	6.8
Solubale Salts		ppm	82	78
Chloride (Cl)		ppm	7	7
Bicabonate (HCO3)		ppm	82	78
ANIONS	SULFUR	ppm	1.38	1.26
	PHOSPHORUS	ppm	0.29	0.37
SOLUBLE CATIONS	CALCIUM	ppm	16.18	16.14
		meq/l	0.81	0.81
	MAGNESIUM	ppm	3.96	2.92
		meq/l	0.33	0.24
	POTASSIUM:	ppm	4.81	5.70
		meq/l	0.12	0.15
	SODIUM	ppm	0.48	0.37
		meq/l	0.02	0.02
PERCENT	Calcium		62.97	66.45
	Magnesium		25.67	20.03
	Potassium		9.72	12.19
	Sodium		1.63	1.33
TRACE ELEMENTS	Boron (p.p.m.)		0.04	0.05
	Iron (p.p.m.)		3.79	2.15
	Manganese (p.p.m.)		0.26	0.13
	Copper (p.p.m.)		0.02	0.02
	Zinc (p.p.m.)		0.07	0.04
	Aluminum (p.p.m.)		1.84	1.07
OTHER				

Table 39: *Saturated Paste Report*

the balance on the standard soil test, the poor field looks to be in better balance. The values on the paste tests are so close I don't think you could run one test twice and get them any closer. These fields have been treated generally with the same products; however, the good field has obviously received more lime and fertilizer over time (the can is more full). Even though the paste test is nearly identical, the good field, with more sitting in reserve, can deliver more nutrients into solution as the roots pull them out. This is not always the case — remember there may or may not be any correlation between the standard test and the paste test, but this example indicates why having both the standard and paste tests are the most informative. I have seen the reverse, where the worse standard test is the best yielding, and because it is the best yielding it tends to pull the levels down faster. Eventually, big yields will lead to depleted soils unless replacement and balance is maintained.

We have already discussed balancing the cations in the nutrient section, but balancing the solubility may seem like a bit more of a daunting task. The paste starting guidelines are listed in Table 40.

Let's discuss each of the parameters and how best to adjust the numbers to our guidelines.

Soil pH

The soil pH on the paste test will be the same as the standard soil test 99 percent of the time. The paste pH may come up a little different from time to time, since the paste pH is read twenty-four hours after the initial saturation. The soil may have some organic acids or carbonates that release over time, causing a fluctuation of a few tenths of a point. You can send in your irrigation water and ask the lab to use that instead of distilled water. Of course, that could have a big impact on the pH reading.

I have had clients do both a distilled water and irrigation water extracts for the paste test, especially for high-valued crops or high-end sports facilities. The distilled water extract is what the plants would see under normal rainfall, but extracting with the irrigation water will indicate the impact your water will have on the movement of nutrients into solution. This allows you to possibly adjust your fertility program ahead of or during irrigation cycles.

Paste guidelines

	Desired Levels TEC<10	Desired Levels TEC>10	Growing Media
Soil pH	6.3-6.6	6.2-6.5	6.2-6.5
Soluble Salts ppm	<500	<600	<450
Anions			
Sulfur ppm	1-3	3-5	1-3
Chlorides ppm	<60	<90	<60
Bicarbonates ppm	<90	<120	<90
Phosphorus ppm	0.3-0.6	0.3-0.6	1.5-3.0
Cations			
Calcium ppm	30-40	20-40	50-70
Magnesium ppm	6-8	4-8	10-14
Potassium ppm	12-15	10-12	20-25
Sodium ppm	<6	<5	<6
Solution %			
Calcium %	60	60	60
Magnesium %	20	20	20
Potassium %	12-15	12-15	12-15
Sodium %	<5	<5	<5
Trace Elements			
Boron ppm	0.05-0.10	0.05-0.10	0.08-0.10
Iron ppm	0.5-1.5	0.5-1.5	0.5-1.5
Manganese ppm	0.07-0.10	0.07-0.10	0.08-0.15
Copper ppm	0.05-0.08	0.05-0.08	0.06-0.10
Zinc ppm	0.07-0.15	0.07-0.15	0.10-0.15
Aluminum ppm	<1.5	<1.5	<1.5

Table 40

Soluble salts, chlorides, and bicarbonates

I will vary these restricting parameters a little based on the buffering capacity of the soil. The higher the exchange capacities, the better the buffering capacity tends to be. These numbers could be a lot higher for plants that are naturally salt tolerant. Larger plants transplanted into soil or growing media will tolerate higher levels better than germinating or small plants. Keep in mind that these numbers are based on a saturated soil condition, so as the soils dry out the concentration of these parameters in the soil solution will even get higher.

Sulfur

Sulfur too is varied a little based on exchange capacity. These values are more as minimum levels, but the levels could go considerably higher. For crops requiring higher nitrogen applications, these minimum levels may have to be raised. If the sulfur numbers get noticeably higher without constantly adding a sulfur source, look to see if there are some drainage issues starting to develop.

Cations

The desired levels for soils tend to be opposite for the above parameters. The low-exchange soils do not hold as much of the cations as the higher-exchange soils, so I prefer to increase the soluble base levels. This also increase the risk of loss during high-rainfall events. The feed rate will not be as constant or long lasting in the low-exchange soils as the higher-exchange-capacity soils. Higher-exchange soils will generally have a higher clay or organic matter content which will hold water, allowing for a more constant feed rate. The growing media tend to have very little buffering capacity in spite of the high organic matter levels, and they are generally growing plants with faster growth rates and shorter growth cycles — hence the higher minimum levels.

Trace Elements

The desired levels of trace elements are similiar for both soil and growing media. This is somewhat due to the lack of comparison tissue data to help better define the ideal levels. Where I have had tissue data for comparison of mainly field crops, these numbers have proved satisfactory.

Chapter 13
Soil Balancing Examples

In this chapter, we will be balancing several soil samples using both organic and commercial products. We will balance several low-exchange-capacity soils using the Sufficiency Level of Available Nutrients (SLAN) approach and several high-exchange-capacity soils using the Albrecht approach, or the Basic Cation Saturation Ratio (BSCR).

1. One of the first things that I look at on a soil report is the exchange capacity. This will immediately indicate what approach I will take in order to balance the soil.
2. The second parameter that I look at is the pH of the sample. This tells me how difficult it will be to balance the soil. Low pH soils are generally easy to fix. High pH levels mean that there are excess levels somewhere and they are harder and take longer to balance, especially if tillage is limiting as well as the drainage. Low rainfall areas are also harder to balance.
3. The third parameter to consider is the base saturation numbers if the soil is a high-exchange-capacity soil. A low-exchange-capacity soil means that I start looking down the report for deficient levels. I will often look at the paste test on these low-exchange soils just to get a feel for the overall solubility before I go back to calculating the materials needed to fix the soil.
4. Once that I know the overall balance of the cations on the high exchange soils, I will decide on a plan to rebalance the cations while trying to keep soil pH within the 6.2-6.5 level. High pH levels will require elemental sulfur or pH-neutral products such as gypsum, KMag, or magnesium sulfate for calcium and

magnesium balance. Low-pH soils can be easily and cheaply balanced using the correct lime — either dolomite or high-calcium lime. Potassium balancing is nearly always accomplished with pH-neutral products. Organic products such as wood ash, layer manure, or compost, if available and cost effective, should always be considered.
5. Sulfur and phosphorus are my next concerns, and they are balanced using the SLAN approach.
6. Lastly, I look at the trace elements and balance them using the SLAN approach. Sometimes, depending on whether there are excessive levels of cations or phosphorus, I will forego adding iron, manganese, copper, or zinc and rely on foliar feeding instead.
7. After I have decided on a plan for balancing the standard soil report, I look at the paste analysis for additional guidance on what products I may want to use. Many times, the standard report has looked quite good but the nutrients are not going into solution, yielding a poor level on the paste test. Then it becomes more of solution balancing process than a colloidal balancing process.

Samples to Balance: Low Cation Exchange Capacity Soils

There will be several samples that we will be balancing in this section. Each example will have a standard test as well as a paste test. All of the samples will be low-exchange soils (<10), which will require balancing using the SLAN approach.

Example 1: Field D is a very low-exchange-capacity soil with pretty much everything wrong. The one bright spot for this soil sample is the great organic matter level. Note the sample depth is 5 inches and the desired levels are adjusted to that depth. The pH is less than 6.5 so we will start with lime recommendations. Magnesium is low and we could supply the magnesium with KMag or Epsom salts, but that would be more expensive than using dolomite lime. Dolomite lime has 11 percent magnesium, so by dividing the deficit (39) by 11 percent we have a limit on the amount of dolomite lime that we need (355 pounds). Since this sampled to 5 inches and Logan sets their magnesium base saturation to 12 percent, of which I personally prefer 15 percent, I would round up this application to 400 pounds. If you are working on pounds per thousand square feet, simply divide the pounds per acre by 43.56 and end up with nearly 10 ponds per 1,000 square feet. Dolomite lime has approximately twice as much calcium as magnesium, so 400 times 21 percent (calcium content of dolomite lime) adds 84 pounds of calcium besides the 44 pounds of magnesium. Subtracting 84 pounds from a 439 pound calcium deficit leaves us with a 355 pound calcium deficit remaining, which we can fix with high-calcium lime. High-calcium lime contains approximately 30 percent calcium. Dividing the 355 pounds calcium deficit by 0.3, we end up needing roughly 1,183 pounds of high-calcium lime besides the 400 pounds of dolomite lime. If you can only put one lime on in the spring and the other lime on in the fall, start with the dolomite since it has both calcium and magnesium. Putting the high-calcium lime on first might induce a magnesium deficiency. Remember that calcium needs to be at the growing point for roots, so incorporating at least 5-6 inches is ideal.

The potassium deficit is 101 pounds, which when corrected might be fine for a pasture, but a vegetable crop would require a much higher rate. Checking the removal rate of the crop you are growing and adding that to the desired level on a low-exchange-capacity soil is a good place to start. Remember that these soils can't hold all this potassium along with the calcium and magnesium, so apply the potassium in the spring ahead of planting and not in the fall. Those growing organically will be using potassium sulfate while commercial growers will probably use muriate of potash. Applying a large amount of muriate of potash on low TEC soils just ahead of planting might create a chloride problem, so applying most of the potash in the fall may be necessary. Either way, both products are very soluble and the anions could be subject to loss over the winter. There is always economics to consider, so adding just the removal rate of potassium along with the lime may be all that you can do in one year. **These soils didn't get out of balance in one year**

and won't get in balance in one year. Whatever the removal rate of potassium, we will divide that number by 0.50 for potassium sulfate (50 percent K) or 0.60 for potassium chloride (60 percent K).

Phosphorus is also very low (41 pounds P_2O_5) and I would like to see a minimum of 200 pounds of P_2O_5 per acre for general crops and higher than that for specialty crops. Correcting a phosphorus problem such as this for organic growers will require chicken manure or some other high-phosphorus source. Composted chicken manure would be very stable, but composted products like chicken manure generally contain only 3-4 percent elemental phosphorus and 7-9 percent P_2O_5. These products are much better for maintaining phosphorus and building organic matter. Adding a chicken manure to supply both the removal as well as the deficit would be the fastest and easiest way to build phosphorus levels.

Let's assume that we want to grow potatoes and want to use chicken manure as a means to correct the phosphorus. Table 4 indicates that we would need approximately 114 pounds of P_2O_5 for the crop and 159 pounds of P_2O_5 to bring the phosphorus to the minimum desired level. This would be a total of 273 pounds of P_2O_5. Table 9 indicates that a ton of chicken manure would contain roughly 48 pounds of P_2O_5. Dividing the sum total of needed phosphorus (273 pounds) by 48 equals 5.7 tons per acre of chicken compost needed. Adding 5 or 6 ton per acre of chicken compost would be very easy to apply with the equipment that we have today. Table 9 also indicates that chicken manure could also contain nearly 33 pounds/ton of both nitrogen and potassium. At a 6-ton rate of chicken manure, we would add 198 pounds of nitrogen and potassium. This would nearly supply the necessary nitrogen and potassium for the potato crop. Using the chicken manure to cover the nitrogen requirements for any crop will eventually result in excess phosphorus in the soil. This happened on the East Coast, where all the broiler production was decades ago. Even now, though the chicken production moved out years ago, their soils still show phosphorus levels of 1,000 pounds or more P_2O_5.

The nitrogen in chicken and turkey manures are primarily in the urease form and not very stable. Fall applying could result in a significant loss or nitrogen before spring planting. Products like compost and rock phosphate could also be an option, but both of these should probably be applied in the fall to start the break down and to get the soil biological life up and running to handle these products. A commercial grower will likely use a MAP (11-52-0) or DAP (18-46-0) product, which is very soluble and best suited for a spring application, in order to maintain solubility in the soil. Even in sands these products will not move more than an inch in the soil. Phosphorus from any source does not move very far in the soil, so incorporation 5-8 inches would be ideal. For the above example, it would take nearly 525 pounds per acre of MAP to supply the phosphorus needed for the potato crop. This would be a lot of soluble phosphorus at any one time and could tie up a lot of zinc. Build-up applications would be best accomplished over a couple of years and incorporated to 6-8 inches. The use of cover crops would help assimilate these soluble products but would also increase the biology for low-soluble products. The paste analysis on a soil with these many problems is probably not necessary at this time.

It is interesting to note that even as low as the phosphorus and the trace elements (excluding iron) are, there is some movement into solubility. This is more than likely the result of the good organic matter in the soil. The soluble phosphorus levels and trace elements will more than likely go down as we add lime. For this reason, we should consider adding trace elements to our fertility program. This should be in the form of dry broadcast and foliar feeding. Building deficit phosphorus levels using manure or rock phosphates does not impact the trace element levels nearly so much since both products will contain a certain amount of trace elements. The pure commercial fertilizers such as MAP or DAP do not contain any trace elements and have a tendency to tie up the positively charged traces. In this example, adding 10 pounds of a 10 percent boron and 15-20 pounds of zinc and manganese sulfate would be a great start for covering crop removal as well as building the trace levels.

Trace elements can really take a bite out of the fertilizer budget. If it is cost prohibitive to go with the dry bulk application, it might be better to try a combination of dry and foliar. Apply with the dry fertilizer application crop removal rates for copper, manganese, and zinc. This

Sample Location			Field D
Sample ID			
Lab Number			32
Sample Depth in inches			5
Total Exchange Capacity (M.E.)			3.97
pH of Soil Sample			5.3
Organic Matter, Percent			6.25
ANIONS	Sulfur	p.p.m.	12
	Mehlich III Phosporous	as (P_2O_5)	41
		lbs / acre	
EXCHANGEABLE CATIONS	CALCIUM lbs / acre	Desired Value	899
		Value Found	460
		Deficit	-439
	MAGNESIUM lbs / acre	Desired Value	166
		Value Found	127
		Deficit	-39
	POTASSIUM lbs / acre	Desired Value	166
		Value Found	65
		Deficit	-101
	SODIUM	lbs / acre	59
BASE SATURATION %	Calcium (60 to 70%)		34.77
	Magnesium (10 to 20%)		16.00
	Potassium (2 to 5%)		2.52
	Sodium (.5 to 3%)		3.86
	Other Bases (Variable)		6.80
	Exchangeable Hydrogen (10 to 15%)		36.00
TRACE ELEMENTS	Boron (p.p.m.)		0.39
	Iron (p.p.m.)		518
	Manganese (p.p.m.)		7
	Copper (p.p.m.)		1.6
	Zinc (p.p.m.)		3.55
	Aluminum (p.p.m.)		281
OTHER	Ammonium (p.p.m.)		1.6
	Nitrate (p.p.m.)		6.8

Example 1: *Soil Report*

Sample Location			Field D
Sample ID			
Lab Number			128585
Water Used			DI
pH			5.3
Soluble Salts		p.p.m.	26
Chloride (Cl)		p.p.m.	8
Bicarbonate (HCO3)		p.p.m.	29
ANIONS	SULFUR	p.p.m.	1.38
	PHOSPOROUS	p.p.m.	0.28
SOLUBLE CATIONS	CALCIUM	p.p.m.	2.96
		meq/l	0.15
	MAGNESIUM	p.p.m.	1.51
		meq/l	0.13
	POTASSIUM	p.p.m.	1.87
		meq/l	0.05
	SODIUM	p.p.m.	1.58
		meq/l	0.07
PERCENT	Calcium (60 to 70%)		37.80
	Magnesium (10 to 20%)		32.24
	Potassium (2 to 5%)		12.39
	Sodium (.5 to 3%)		17.56
TRACE ELEMENTS	Boron (p.p.m.)		0.03
	Iron (p.p.m.)		2.05
	Manganese (p.p.m.)		0.05
	Copper (p.p.m.)		0.03
	Zinc (p.p.m.)		0.03
	Aluminum (p.p.m.)		1.12
OTHER			

Example 1A: *Saturated Paste Report*

would amount to 2-3 ounds of copper sulfate and 5 pounds of manganese and zinc sulfate. Boron 10 percent would be applied at the full rate of 10-15 pounds/acre for most crops. Boron is leachable and will be a yearly application. There is so much iron in the soil that I would let it go unless I'm in sands or in irrigation situations where there are high bicarbonates in the water. In those cases, I would foliar feed iron along with the other traces at peak demand times.

Example 2: Lawn E is another low-exchange-capacity soil that looks much better than the previous test. Even though this is a sample from a lawn, we will balance it much the same way as if it were going to general crops or vegetables. I want to start with the standard test and end up with the paste test. Oftentimes the standard test looks good on the low-exchange soils but comes up short in the paste test.

The pH is good and the calcium is a little bit on the high side. The magnesium is great, but potassium is very short, as it is on many high-end lawns where the grass clippings are removed. So, potassium is the first adjustment that we need on this report. For turf, I would like to have the potassium around 250 pounds/acre for these light soils. Potassium will help in disease resistance, drought, cold tolerance, and movement of carbohydrates, whether this is a yard, a vegetable crop, or field corn. If this was vegetables, I would push the potassium up to 400-500 pounds/acre. This soil is 119 pounds deficit and I would like to have a 50 pound cushion, so we need to add roughly 170 pounds of potassium. I would use potassium sulfate whether or not this was a yard that was in an organic program. The sulfur is low and this would fix both potassium and the sulfur problems. Dividing the 170 pound deficit by 0.5 (percent potassium in potassium sulfate) we would need 340 pounds of product. This would be nearly 8 pounds per 1,000 square feet and could be applied all at once early spring or split applied in two applications. Core aerating prior to at least one application would be ideal. In a cropping situation, I would make one application and work in 6-8 inches. Potassium chloride would be used in most commercial operations due to the low cost. Dividing 170 pound deficit by 0.6 would mean that nearly 280 pounds would be needed to supply the desired amount of potassium.

Ammonium sulfate would have to be used as part of the nitrogen source in order to cover the sulfur deficiency. Boron, manganese, and zinc are low on the standard test, but I like to check the paste test at this point to see if they are also low. Boron and zinc is low on the paste test but manganese is good. I would still add manganese in the fertility program since the standard test is so very low. This sample was collected in February when the grass is not growing and I would be concerned if the feed rate would be fast enough during the growing season. Ten pounds of boron, 15 of manganese and 20 pounds of zinc would cover the turf, but I would probably jump each of the products by 5 pounds for any specialty crop production. Just for the cost concerns for general crops, I would leave the 10 pounds of boron but cut the manganese and zinc by 10 pounds and try to put a trace mix in the starter and/or foliar feed.

The corrections made from the standard test would also cover the potassium, sulfur, and trace element shortages on the paste test. The paste test indicates some magnesium and phosphorus shortages. Here is where I would go back and lower the previous potassium applications and add in a little KMag to boost the soluble magnesium. We just want to increase the solution magnesium and not the colloidal magnesium. I would try to accomplish this by supplying the crop removal rate of magnesium (Table 19) with KMag. For example, 200 bushels of corn would remove approximately 18 pounds of magnesium. This would require a broadcast application of nearly 160 pounds of KMag and a reduction of 60 pounds of potassium chloride, or 72 pounds of potassium sulfate.

Quite often, phosphorus is below the detection limit on the paste analysis (Example 2A) when there are good or even high numbers on the standard test (Example 2). Do not dismiss this situation, especially when starting or growing plants under cold and wet conditions. Phosphorus availability is strongly influenced by temperature and soil biology. Even with soil levels of 649 pounds P_2O_5 like we have in Example 2, it would be a good idea to supply some soluble phosphorus for a few weeks until soil temperatures rise and the soil biology can get ramped up. This could be accomplished by using an ortho-phosphorus product in the transplant solution or a small amount of phosphorus in a row starter. Winter wheat planted late in the fall that failed to tiller properly should have some soluble phosphorus in the green-up nitrogen application. Just 4-5 gallons of a 10-34-0 poly phosphate or 25-50 pounds of a MAP or DAP dry fertilizer would help get you over the hump. Adequate levels of potassium, magnesium, and zinc all improve phosphorus uptake. I would try and not apply more than 10-20 percent of the phosphorus needs to a crop when I have levels over 500 pounds P_2O_5. For general agricultural crops, I would drop that level to 200-250 pounds P_2O_5, depending upon organic matter levels and overall soil conditions. The best way to develop your minimum levels is through some detailed tissue analysis. I have seen phosphorus levels between a 1,000 pounds and 2,000 pounds per acre of P_2O_5 from manure applications, and still crop production was very good. However, pushing the phosphorus levels up higher than these minimum levels is a waste of money and poses an environmental concern. One of the best ways to maximize phosphorus availability is by enhancing the biology in the soil. Adding sugars, humic acids, and compost teas to starters and transplant solutions is an easy way to kick off the biology in the spring. Cover cropping is another excellent way to maintain a high level of biological activity in the soil.

Sample Location			Lawn E
Sample ID			
Lab Number			44
Sample Depth in inches			6
Total Exchange Capacity (M.E.)			5.76
pH of Soil Sample			6.7
Organic Matter (percent)			4.48
ANIONS	SULFUR:	p.p.m.	13
	Mehlich III Phosphorous:	as (P_2O_5) lbs / acre	649
EXCHANGEABLE CATIONS	CALCIUM: lbs / acre	Desired Value	1565
		Value Found	1666
		Deficit	
	MAGNESIUM: lbs / acre	Desired Value	200
		Value Found	209
		Deficit	
	POTASSIUM: lbs / acre	Desired Value	200
		Value Found	81
		Deficit	-119
	SODIUM:	lbs / acre	40
BASE SATURATION %	Calcium (60 to 70%)		72.35
	Magnesium (10 to 20%)		15.13
	Potassium (2 to 5%)		1.80
	Sodium (0.5 to 3%)		1.51
	Other Bases (Variable)		4.70
	Exchangeable Hydrogen (10 to 15%)		4.50
TRACE ELEMENTS	Boron (p.p.m.)		0.51
	Iron (p.p.m.)		241
	Manganese (p.p.m.)		8
	Copper (p.p.m.)		10.64
	Zinc (p.p.m.)		2.48
	Aluminum (p.p.m.)		1052
OTHER			

Example 2: *Soil Report*

Sample Location			Lawn E
Sample ID			
Lab Number			127949
Water Used			DI
pH			6.7
Solubale Salts		ppm	108
Chloride (Cl)		ppm	10
Bicabonate (HCO3)		ppm	105
ANIONS	SULFUR	ppm	1.75
	PHOSPHORUS	ppm	< 0.03
SOLUBLE CATIONS	CALCIUM	ppm	18.89
		meq/l	0.94
	MAGNESIUM	ppm	5.30
		meq/l	0.44
	POTASSIUM:	ppm	3.48
		meq/l	0.09
	SODIUM	ppm	5.17
		meq/l	0.22
PERCENT	Calcium		55.52
	Magnesium		25.94
	Potassium		5.31
	Sodium		13.22
TRACE ELEMENTS	Boron (p.p.m.)		0.03
	Iron (p.p.m.)		1.17
	Manganese (p.p.m.)		0.07
	Copper (p.p.m.)		0.04
	Zinc (p.p.m.)		< 0.02
	Aluminum (p.p.m.)		8.15
OTHER			

Example 2A: *Saturated Paste Report*

The third and last example of balancing a low exchange capacity soil is found in Example 3 and 3A.

Sample Location			A5
Sample ID			
Lab Number			162
Sample Depth in inches			6
Total Exchange Capacity (M.E.)			8.47
pH of Soil Sample			6.7
Organic Matter (percent)			3.75
ANIONS	SULFUR:	p.p.m.	16
	Mehlich III Phosphorous:	as (P_2O_5) lbs / acre	645
EXCHANGEABLE CATIONS	CALCIUM: lbs / acre	Desired Value	2304
		Value Found	2408
		Deficit	
	MAGNESIUM: lbs / acre	Desired Value	244
		Value Found	268
		Deficit	
	POTASSIUM: lbs / acre	Desired Value	264
		Value Found	373
		Deficit	
	SODIUM:	lbs / acre	37
BASE SATURATION %	Calcium (60 to 70%)		71.05
	Magnesium (10 to 20%)		13.18
	Potassium (2 to 5%)		5.64
	Sodium (0.5 to 3%)		0.94
	Other Bases (Variable)		4.70
	Exchangeable Hydrogen (10 to 15%)		4.50
TRACE ELEMENTS	Boron (p.p.m.)		0.67
	Iron (p.p.m.)		337
	Manganese (p.p.m.)		70
	Copper (p.p.m.)		1.91
	Zinc (p.p.m.)		6.44
	Aluminum (p.p.m.)		385
OTHER	Ammonium (p.p.m.)		0.9
	Nitrate (p.p.m.)		2.8

Example 3: *Soil Report*

Sample Location			A5
Sample ID			
Lab Number			127666
Water Used			DI
pH			6.7
Solubale Salts		ppm	97
Chloride (Cl)		ppm	7
Bicabonate (HCO3)		ppm	84
ANIONS	**SULFUR**	ppm	3.36
	PHOSPHORUS	ppm	0.7
SOLUBLE CATIONS	**CALCIUM**	ppm	16.12
		meq/l	0.81
	MAGNESIUM	ppm	4.25
		meq/l	0.35
	POTASSIUM:	ppm	10.97
		meq/l	0.28
	SODIUM	ppm	1.52
		meq/l	0.07
PERCENT	Calcium		53.34
	Magnesium		23.43
	Potassium		18.86
	Sodium		4.38
TRACE ELEMENTS	Boron (p.p.m.)		0.09
	Iron (p.p.m.)		3.54
	Manganese (p.p.m.)		0.07
	Copper (p.p.m.)		0.02
	Zinc (p.p.m.)		0.04
	Aluminum (p.p.m.)		3.66
OTHER			

Example 3A: *Saturated Paste Report*

Here is a standard soil with very good balance, but for whatever reason the solubility needs some improvement. Remember that there may or may not be a correlation between the standard test and the paste test. That is why it is a good idea to always run a paste analysis on at least some of your soil samples, if doing multiple samplings. This practice will give you at least a basic idea of what is truly available to the plants in the soil. If I were forced to choose between a tissue analysis and a paste analysis, I would choose the paste analysis since it best reflects what is available to the plant.

The standard test for Example 3 is very good, with sulfur being the only major nutrient on the low side. The trace element results also show boron, copper, and zinc on the low side. The zinc is better than I typically see, but due the high phosphorus, it may be restricted for plant uptake.

The paste test is a bit the opposite of the standard test. The sulfur is low like the standard test, but calcium and magnesium are low in spite of good numbers on the other test. The trace elements copper and zinc are low, but boron appears to be quite adequate. Copper is not really dependable on the paste test since it is held so tightly on the clays and easily bound by organic complexes. Since the results of copper analysis are low on both tests, I would be inclined to add 5 pounds of copper sulfate to my broadcast applications and include copper in my foliar sprays if that is in the plans for the crop. Boron is good on the paste test but a little low on the standard test. A typical boron application for me, without a paste test, might be 10 pounds for corn and 15 pounds for specialty crops, but with this paste analysis I would feel comfortable dropping off 5 pounds of boron per acre. At 0.75 cents per pound that would be a savings of $3.75 per acre, and at 10 acres the cost of the analysis would easily be covered. Zinc is low in the paste test but above the detection limit of 0.02. This indicates there is some zinc solubility; so I would apply a maintenance level of 5 pounds through the dry fertilizer program and foliar apply zinc during the growing season. If foliar is not an option, then banding in the row would be the next best alternative. It is questionable as to how much of the broadcast zinc will be used by the plants, especially with the high phosphorus, however a broadcast zinc sulfate will be one of the cheapest ways to maintain the level.

Calcium and magnesium are low and the pH is a little on the high side, so sulfated forms of both calcium and magnesium would be the products of choice. These are pH-neutral products and will also supply much-needed sulfur. To raise the calcium to 40 ppm would require roughly 25 ppm of a soluble calcium source (gypsum) to be added to the soil. Assuming that a 6-inch soil slice weighs 2 million pounds, then 25 x 2 = 50 pounds of soluble calcium would be needed. The calcium percentage of gypsum is 23 percent. Dividing 50 by 0.23 means that 217 pounds of gypsum would have to be added to the soil. Raising the magnesium 4 ppm based on similar assumptions means that we would have to add 8 pounds of magnesium. The magnesium content of epsom salts is 17 percent. Dividing 8 by 0.17 equals 47 pounds of epsom salt that would have to be added to the soil.

These calculations are made to bring the soil paste balance to our ideal numbers, but what happens to that balance as the crops grow and remove nutrients or rain begins to perk through the soil, taking soluble nutrients with it? Those ideal numbers are going to decrease, so enough extra nutrients must be supplied to get the plants through their high-demand periods. Calcium and magnesium are on high demand all the way through vegetative growth and into flowering. Once the plant has reached maximum size, the demand for calcium rapidly decreases. Tissue sampling at this point would indicate whether enough soluble calcium and magnesium were applied. Weather could play a big role on uptake — it should be considered in the overall evaluation. The need for potassium starts a few weeks after emergence all the way through the reproductive phase and on into the fruit development stage. A late tissue sample of the lower leaves would indicate the potassium status for the crop. By studying the roles of each of the nutrients in the plant, it is possible to predict to some degree the plant's demand for nutrients and where best to do the tissue sampling.

Traces are the most difficult because of the sheer number of chemical processes that they are involved in, but traces are just that — nutrients needed in minute amounts. Applying various ores or rock dusts along with banding and foliar feeding trace elements can fulfill most, if not all of the plant's needs. Just don't get caught up with some silver-bullet product without truly understanding the major and secondary nutrient needs.

Samples to Balance: High Cation Exchange Capacity Soils

In this section, we are going to balance high-exchange-capacity soils using the Basic Cation Saturation Ratio, also referred to as the Albrecht philosophy. The BCSR is only used to balance the basic cations. All anions and trace elements are balanced using the Sufficiency Level of Available Nutrients, commonly called the SLAN approach. The goal for balancing the cations is to improve the flocculation of the clays in order to improve soil structure and, in turn improve infiltration, percolation, and aeration. The soil biology will greatly benefit from this action, as well as crop yield and quality.

Many labs report the soil test data as parts per million. The first report that we will balance will be in both ppm and pounds per acre. I prefer pounds/ac. since that is how we buy nutrients. We apply them by the acre or by pounds/ft^2.

In Example 4 the cations are in parts per million. I like to convert them to pounds per acre to truly understand what I'm dealing with. Converting ppm to pounds per acre is very simple. Take the sample depth and divide by three. This assumes that every 3 inches of soil weighs one million pounds. Therefore, in Example 4, the sample depth is listed at 6 inches and so dividing it by three equals two. Multiplying the desired cation values, values found, and the deficits by two will give you the pounds per acre.

Phosphorus listed as P_2O_5 could be converted to pounds per acre the same as the cations. Remember that when we buy phosphorus, we are buying it as P_2O_5 and not P. Elemental phosphorus does not exist in nature. It exists as in an oxide form as P_2O_5. Many labs report phosphorus as elemental P. To convert elemental P to P_2O_5, the conversion factor is 2.29. The elemental P value in Example 4 is 30 ppm; therefore, by multiplying 30 by the conversion factor of 2.29, we end up with 68.7

ppm of P_2O_5. Multiplying 68.7 ppm times two (conversion factor to get to pounds/ac. based on depth) we end up with 138 pounds of P_2O_5 per acre. If our desired level is 200 pounds/ac., then we are 62.6 pounds of P_2O_5 short of our goal. Rock phosphate is 25 percent P_2O_5, so dividing 62.6 by 0.25 tells us that it will take nearly 250 pounds of rock to balance the phosphorus. MAP or 11-52-0 is 52 percent P_2O_5 and it would take only 120 pounds of a commercial-grade product to balance the phosphorus. Example 4A is the same as Example 4 only calculated in pounds per acre.

The calcium and magnesium balance is poor on this soil. The 31 percent magnesium saturation is quite high. This situation tends to make the soils tight and sticky. Grass control problems will be elevated. Percolation of water will be slow, tending to increased anaerobic conditions. Often this high magnesium saturation will lead to low solubility numbers — for magnesium as well as the other cations. In reviewing Example 4B this is shown to be the case. When looking at a feed analysis or tissue analysis, the magnesium will show up low. Intuitively one would think that we should add more magnesium, but the opposite is true. In this case, since the pH is high, gypsum would be the product of choice. The calcium deficit is 330 pounds/acre, so dividing 330 by the percent of calcium in gypsum (23 percent) we can calculate that 1,435 pounds of gypsum would be needed to satisfy the calcium deficit. Assuming that we could remove only the magnesium and replace it with calcium, the balance of the soil would still only be 68 percent and the magnesium 23 percent. This is better, but we are working to get to 15 percent magnesium, which will require adding additional gypsum. Adding 1,500 pounds of gypsum this year and following up with another 1,500 pounds of gypsum the following year will really begin to bring this soil into balance. There is just one catch. There must be an avenue for the magnesium to be moved out of the root zone. Any compaction layers will slow percolating water down and allow the magnesium to be deposited on top of the compaction zone and then be brought back up with the next tillage pass.

Another problem that is very likely to occur is the presence of free magnesium carbonate in the soil. Soils over-limed with dolomite lime will more than likely have free carbonates in the soil profile. Free carbonates simply mean that there is raw undissolved lime in the soil. Once the soil pH levels are above 7, the solubility of lime drops to 10 percent or less. Coarser lime will even take longer to dissolve. Until this excess magnesium is cleared out of the root zone, high levels of magnesium will persist on the soil test. Another way to think about this situation is to imagine a glass of water in which sugar has been added until there is a thick layer of sugar on the bottom of the glass. The solution is saturated with sugar and can hold no more, hence the layer on the bottom of the glass. If you pour some of the sugar water out and replace it with clean water, what happens to the concentration of the sugar in the water? Nothing, as long as there is sugar on the bottom of the glass to dissolve. Once all the sugar is dissolved and you pour water out and replace it with clean water, the concentration of sugar in the water rapidly drops. This is the same with magnesium as long as free magnesium carbonates are present in the soil. I have seen it take a decade or longer to clear out excess magnesium, so don't get disappointed when it takes longer than you think to balance your soil. The higher the total exchange capacities, the longer it takes to balance your soil, but the bright side is that it will stay in balance longer.

I don't convert the trace elements to pounds per acre, because the traces are either in the acceptable range or not. If they are not in the acceptable range, I generally apply a set amount that is above removal rate and financially tolerable. If the traces are really low, I prefer to soil-apply something above removal rate. The boron level in Example 4 is low on the standard test and below the detection limit on the paste. In this case, I would soil-apply a 15 pounds/ac. rate of a 10 percent boron. If boron on the standard test was good, but low in the paste test, I would still apply at least 5-10 pounds/ac. depending upon the crop and its overall demand and response to boron. Iron and manganese is good on the standard test and iron is good in the paste test, but manganese is low in the paste and ideally would best be corrected with a foliar application during the vegetative cycle. Zinc and copper are both low in each test and a soil application and foliar application would be the ideal scenario. I would soil-apply 5 pounds/ac. of copper sulfate and 10 pounds/ac. of zinc sulfate and foliar feed during vegetative growth.

Sample Location	Tilled
Sample ID	Area
Lab Number	24
Sample Depth in inches	6
Total Exchange Capacity (M.E.)	10.87
pH of Soil Sample	7.1
Organic Matter (percent)	2.60

ANIONS	SULFUR:	p.p.m.	7
	Mehlich III Phosphorous:	ppm	30
EXCHANGEABLE CATIONS	CALCIUM: ppm	Desired Value Value Found Deficit	1478 1313 -164
	MAGNESIUM: ppm	Desired Value Value Found Deficit	156 410
	POTASSIUM: ppm	Desired Value Value Found Deficit	169 129 -40
	SODIUM:	ppm	21
BASE SATURATION %	Calcium (60 to 70%)		60.41
	Magnesium (10 to 20%)		31.46
	Potassium (2 to 5%)		3.03
	Sodium (0.5 to 3%)		0.82
	Other Bases (Variable)		4.30
	Exchangeable Hydrogen (10 to 15%)		0.00
TRACE ELEMENTS	Boron (p.p.m.)		0.37
	Iron (p.p.m.)		172
	Manganese (p.p.m.)		75
	Copper (p.p.m.)		1.94
	Zinc (p.p.m.)		1.43
	Aluminum (p.p.m.)		507
OTHER	Ammonium (p.p.m.)		1.7
	Nitrate (p.p.m.)		2.9

Example 4: *Soil Report*

Sample Location	Tilled
Sample ID	Area
Lab Number	24
Sample Depth in inches	6
Total Exchange Capacity (M.E.)	10.87
pH of Soil Sample	7.1
Organic Matter (percent)	2.60
ANIONS — SULFUR: p.p.m.	7
Mehlich III Phosphorous: as (P_2O_5) lbs / acre	138
EXCHANGEABLE CATIONS — CALCIUM: lbs / acre — Desired Value	2957
Value Found	2627
Deficit	-330
MAGNESIUM: lbs / acre — Desired Value	313
Value Found	821
Deficit	
POTASSIUM: lbs / acre — Desired Value	339
Value Found	257
Deficit	-82
SODIUM: lbs / acre	41
BASE SATURATION % — Calcium (60 to 70%)	60.41
Magnesium (10 to 20%)	31.46
Potassium (2 to 5%)	3.03
Sodium (0.5 to 3%)	0.82
Other Bases (Variable)	4.30
Exchangeable Hydrogen (10 to 15%)	0.00
TRACE ELEMENTS — Boron (p.p.m.)	0.37
Iron (p.p.m.)	172
Manganese (p.p.m.)	75
Copper (p.p.m.)	1.94
Zinc (p.p.m.)	1.43
Aluminum (p.p.m.)	507
OTHER — Ammonium (p.p.m.)	1.7
Nitrate (p.p.m.)	2.9

Example 4A: *Soil Report*

Sample Location			Tilled
Sample ID			Area
Lab Number			127567
Water Used			DI
pH			7.1
Solubale Salts		ppm	69
Chloride (Cl)		ppm	7
Bicabonate (HCO3)		ppm	83
ANIONS	SULFUR	ppm	0.48
	PHOSPHORUS	ppm	0.1
SOLUBLE CATIONS	CALCIUM	ppm	12.03
		meq/l	0.60
	MAGNESIUM	ppm	4.90
		meq/l	0.41
	POTASSIUM:	ppm	2.03
		meq/l	0.05
	SODIUM	ppm	0.49
		meq/l	0.02
PERCENT	Calcium		55.51
	Magnesium		37.68
	Potassium		4.86
	Sodium		1.95
TRACE ELEMENTS	Boron (p.p.m.)		< 0.02
	Iron (p.p.m.)		1.56
	Manganese (p.p.m.)		0.04
	Copper (p.p.m.)		< 0.02
	Zinc (p.p.m.)		< 0.02
	Aluminum (p.p.m.)		1.18
OTHER			

Example 4B: *Saturated Paste Report*

The next high-exchange-capacity sample is somewhat similar to the previous sample as far as the number of deficiencies and high magnesium, but this sample has a pH of 6.1 instead of 7.1.

	Sample Location		C1
	Sample ID		
	Lab Number		7
	Sample Depth in inches		6
	Total Exchange Capacity (M.E.)		12.28
	pH of Soil Sample		6.1
	Organic Matter (percent)		5.31
ANIONS	SULFUR:	p.p.m.	6
	Mehlich III Phosphorous:	as (P_2O_5) lbs / acre	56
EXCHANGEABLE CATIONS	CALCIUM: lbs / acre	Desired Value / Value Found / Deficit	3339 / 2688 / -651
	MAGNESIUM: lbs / acre	Desired Value / Value Found / Deficit	353 / 735 /
	POTASSIUM: lbs / acre	Desired Value / Value Found / Deficit	383 / 79 / -304
	SODIUM:	lbs / acre	45
BASE SATURATION %	Calcium (60 to 70%)		54.73
	Magnesium (10 to 20%)		24.94
	Potassium (2 to 5%)		0.82
	Sodium (0.5 to 3%)		0.80
	Other Bases (Variable)		5.20
	Exchangeable Hydrogen (10 to 15%)		13.50
TRACE ELEMENTS	Boron (p.p.m.)		0.44
	Iron (p.p.m.)		337
	Manganese (p.p.m.)		30
	Copper (p.p.m.)		3.72
	Zinc (p.p.m.)		2.13
	Aluminum (p.p.m.)		472
OTHER	Ammonium (p.p.m.)		1.1
	Nitrate (p.p.m.)		1.3

Example 5: *Soil Report*

			C1
Sample Location			
Sample ID			
Lab Number			127896
Water Used			DI
pH			6.1
Solubale Salts		ppm	59
Chloride (Cl)		ppm	5
Bicabonate (HCO3)		ppm	74
ANIONS	SULFUR	ppm	0.54
	PHOSPHORUS	ppm	0.31
SOLUBLE CATIONS	CALCIUM	ppm	8.10
		meq/l	0.40
	MAGNESIUM	ppm	4.71
		meq/l	0.39
	POTASSIUM:	ppm	1.19
		meq/l	0.03
	SODIUM	ppm	2.36
		meq/l	0.10
PERCENT	Calcium		43.50
	Magnesium		42.17
	Potassium		3.31
	Sodium		11.02
TRACE ELEMENTS	Boron (p.p.m.)		0.05
	Iron (p.p.m.)		0.77
	Manganese (p.p.m.)		0.25
	Copper (p.p.m.)		< 0.02
	Zinc (p.p.m.)		< 0.02
	Aluminum (p.p.m.)		0.7
OTHER			

Example 5B: *Saturated Paste Report*

This sample could very well have had a pH level around 6.5 if the potassium was not so low. The soil pH is a product of all the cations, even though we typically think of the pH being effected by calcium and magnesium. Just adding potassium would bring the pH up, but the soil is still out of balance in many ways. The calcium and potassium are low, but the pH is not that far out of the desired range. The magnesium is high, so my choice of calcium would be a combination of gypsum and high-calcium lime. If forced to choose only one product, I would use the gypsum. Gypsum by itself would take nearly 2,800 pounds (651/0.23). A combination of half gypsum and half high-calcium lime would be cheaper and probably a little longer lasting, but a little more work. Twelve hundred pounds each of high-calcium lime and gypsum should nearly satisfy the calcium deficiency and the low sulfur levels on the test. The potassium is very low and the high magnesium could easily interfere with what little potassium is available. Adding the calcium products without adding potassium would only exacerbate this issue. It would take nearly 600 pounds of potassium sulfate or 500 pounds of potassium chloride. I would prefer to correct this deficiency with two applications. One application of potassium chloride would put too much chloride in the soil solution possibly hurting seed germination. It would be best to put the bulk on in the fall where the chlorides could leach out with the winter and spring rains. The more arid climates might want to lean toward a combination of potassium sources or just potassium sulfate.

The phosphorus is very low on the standard test. As low as it is, there is still some phosphorus in solution on the paste test. Since the pH is low, rock phosphate would a viable option. Six hundred pounds of rock phosphate would bring you to the 200-pound desired level for general crops. Adding an additional 400-600 pounds would bring up the levels, which would be more conducive for vegetables. A 400-pound application of MAP (11-52-0) would also add enough phosphorus for general crops, but due to its purity and high solubility, it would tie up calcium and zinc. Both potassium and phosphorus are low, so splitting the application over time would be preferred. A combination of rock phosphate and the bulk of potassium in the fall, as well as MAP and a little potassium in the spring, would avoid hurting zinc and calcium levels. Fall applying the calcium lime and spring applying the gypsum would be the ideal way to handle the calcium products.

Boron, copper, and zinc are low in both the standard and paste test. Blending boron and copper with the P and K would work, but zinc would rapidly tie up with the phosphorus. Applying 5 pounds of zinc sulfate as a maintenance level with the P and K would be acceptable; however, some additional zinc in the row starter or foliar would be ideal.

Example 6 is a soil that looks pretty good on the standard test, but is rather out of balance on the paste test. This soil would grow good general crops, where the demand is not as great as it is for specialty crops.

The standard test has a high level of phosphorus, which basically means that it won't be necessary to purchase phosphorus additives for quite some time. The magnesium is a little high, but the organic matter of 12.37 percent will tend to override a compaction or tightness issue. Basically, this soil has a shortage of sulfur and a little potassium shortage, along with low boron and copper levels. A potassium sulfate application of 275 pounds (138/.50) would fix both the sulfur and potassium issues. Adding 10-15 pounds of a 10 percent boron product should add nearly 0.5-0.75 ppm to the already 0.57 ppm in the soil, taking the total into the desired range of 1.0-1.5 ppm of boron. I would keep the soil applications of copper to around 5 pounds/ac. and not more than 10 pounds/ac. of a 25 percent copper sulfate. Copper sulfate is quite effective at killing algae in ponds, so move copper soil levels slow and easy. Use tissue analysis to guide you to the appropriate levels. Remember that copper is primarily picked up by direct root intercept, so anything that affects root mass will affect copper levels in the tissue.

This would be a good sample to figure out how much nitrogen would need to be added to grow a crop. Let's assume that we are growing mixed vegetables. In general, vegetables will need 90-120 pounds of nitrogen per acre. For our purposes, we are planning for top production, so we are going to use 120 pounds of nitrogen as our desired number. The next thing to do is to calculate the estimated nitrogen release from the organic matter. The fifth row in Table 3 gives us the formula for calculating estimated nitrogen release for soils between 10 and 20 percent organic matter.

Sample Location			C1
Sample ID			
Lab Number			7
Sample Depth in inches			6
Total Exchange Capacity (M.E.)			12.28
pH of Soil Sample			6.1
Organic Matter (percent)			5.31
ANIONS	SULFUR:	p.p.m.	6
	Mehlich III Phosphorous:	as (P_2O_5) lbs / acre	56
EXCHANGEABLE CATIONS	CALCIUM: lbs / acre	Desired Value Value Found Deficit	3339 2688 -651
	MAGNESIUM: lbs / acre	Desired Value Value Found Deficit	353 735
	POTASSIUM: lbs / acre	Desired Value Value Found Deficit	383 79 -304
	SODIUM:	lbs / acre	45
BASE SATURATION %	Calcium (60 to 70%)		54.73
	Magnesium (10 to 20%)		24.94
	Potassium (2 to 5%)		0.82
	Sodium (0.5 to 3%)		0.80
	Other Bases (Variable)		5.20
	Exchangeable Hydrogen (10 to 15%)		13.50
TRACE ELEMENTS	Boron (p.p.m.)		0.44
	Iron (p.p.m.)		337
	Manganese (p.p.m.)		30
	Copper (p.p.m.)		3.72
	Zinc (p.p.m.)		2.13
	Aluminum (p.p.m.)		472
OTHER	Ammonium (p.p.m.)		1.1
	Nitrate (p.p.m.)		1.3

Example 6A: *Soil Report*

Sample Location			Top
Sample ID			Terrace
Lab Number			128567
Water Used			DI
pH			6.8
Solubale Salts		ppm	116
Chloride (Cl)		ppm	7
Bicabonate (HCO3)		ppm	123
ANIONS	SULFUR	ppm	0.7
	PHOSPHORUS	ppm	0.67
SOLUBLE CATIONS	CALCIUM	ppm	18.88
		meq/l	0.94
	MAGNESIUM	ppm	7.87
		meq/l	0.66
	POTASSIUM:	ppm	7.05
		meq/l	0.18
	SODIUM	ppm	0.83
		meq/l	0.04
PERCENT	Calcium		51.90
	Magnesium		36.04
	Potassium		10.06
	Sodium		1.99
TRACE ELEMENTS	Boron (p.p.m.)		0.03
	Iron (p.p.m.)		1.72
	Manganese (p.p.m.)		0.22
	Copper (p.p.m.)		< 0.02
	Zinc (p.p.m.)		0.02
	Aluminum (p.p.m.)		0.61
OTHER			

Example 6B: *Saturated Paste Report*

ENR=125 + [(O.M.-10.0) X 5] if O.M. is > 10% and < 20%

Therefore, 125 + [(12.37.-10.0) X 5]
        125 + [(2.37) X 5]
        125 + 11.85 = 136.85 estimated pounds of nitrogen to be released over the course of the growing season.

From this calculation, it is easy to see that having 10-12 percent organic matter will supply virtually all the nitrogen needed for the growing season. There is a possible exception to this scenario. Short-season crops like lettuce planted early in cool soils may need a little starter nitrogen since the microbes are not at peak numbers to help release organic nitrogen. A lot of excess nitrogen does not hurt crops like corn, but cabbage heads may split and tomatoes may develop excessive growth and cracks in the fruit. Excessive nitrogen can also cause leafy greens to be bitter.

Example 7 is a situation that I commonly see in gardens and hoop houses, which have excessive levels of many nutrients. These situations are very hard to correct, especially in hoop houses. Once the levels are excessive, about the only thing to do is to try and leach the nutrients or dilute with deeper tillage or by adding additional soil.

Excess cations can be leached out of the soil, but phosphorus is virtually impossible. This leaching process assumes that the internal drainage is very good. This soil is very high in phosphorus, magnesium, and potassium. The pH is just a little high considering the level of excess magnesium and potassium. The high level of organic matter probably helps in buffering this situation. The organic matter level will also help the soil in resisting the compaction effect from the excessive magnesium. The calcium is quite low on the standard test, but one look at the paste test and you can see that it is not too bad. The overall balance on the paste test could be improved by adding some calcium to offset the high magnesium and potassium. Since the sulfur is low and we want to increase the calcium, gypsum would be the product of choice. The question becomes, "How much?" If the soil structure is fairly decent,

Sample Location	Flower
Sample ID	Garden
Lab Number	164
Sample Depth in inches	6
Total Exchange Capacity (M.E.)	18.89
pH of Soil Sample	7.1
Organic Matter (percent)	19.53

ANIONS	SULFUR:	p.p.m.	16
	Mehlich III Phosphorous:	as (P_2O_5) lbs / acre	1202
EXCHANGEABLE CATIONS	CALCIUM: lbs / acre	Desired Value Value Found Deficit	5137 3915 -1222
	MAGNESIUM: lbs / acre	Desired Value Value Found Deficit	543 1543
	POTASSIUM: lbs / acre	Desired Value Value Found Deficit	589 1310
	SODIUM:	lbs / acre	82
BASE SATURATION %	Calcium (60 to 70%)		51.82
	Magnesium (10 to 20%)		34.04
	Potassium (2 to 5%)		8.89
	Sodium (0.5 to 3%)		0.95
	Other Bases (Variable)		4.30
	Exchangeable Hydrogen (10 to 15%)		0.00
TRACE ELEMENTS	Boron (p.p.m.)		1.83
	Iron (p.p.m.)		106
	Manganese (p.p.m.)		90
	Copper (p.p.m.)		3.81
	Zinc (p.p.m.)		17.2
	Aluminum (p.p.m.)		401
OTHER	Ammonium (p.p.m.)		2.4
	Nitrate (p.p.m.)		1

Example 7A: *Soil Report*

Sample Location			Flower
Sample ID			Garden
Lab Number			128545
Water Used			DI
pH			7.1
Solubale Salts		ppm	235
Chloride (Cl)		ppm	7
Bicabonate (HCO3)		ppm	227
ANIONS	**SULFUR**	ppm	1.46
	PHOSPHORUS	ppm	2.46
SOLUBLE CATIONS	**CALCIUM**	ppm	24.61
		meq/l	1.23
	MAGNESIUM	ppm	17.90
		meq/l	1.49
	POTASSIUM:	ppm	33.57
		meq/l	0.87
	SODIUM	ppm	1.94
		meq/l	0.08
PERCENT	Calcium		33.45
	Magnesium		40.55
	Potassium		23.70
	Sodium		2.29
TRACE ELEMENTS	Boron (p.p.m.)		0.31
	Iron (p.p.m.)		4.58
	Manganese (p.p.m.)		0.49
	Copper (p.p.m.)		0.02
	Zinc (p.p.m.)		0.04
	Aluminum (p.p.m.)		1.2
OTHER			

Example 7B: *Saturated Paste Report*

which I would think it would be with over 19 percent organic matter, I would consider just balancing the calcium in the paste test. I like to see the calcium to magnesium ratio in the paste test to be 5:1. This would mean raising the calcium from 24.6 ppm to 89.5 ppm (17.9 x 5) or adding 65 ppm of calcium. Sixty-five parts per million in a 6" furrow slice would equal 130 pounds of calcium (65 x 2,000,000). Gypsum is 23 percent calcium, so by dividing 130 by 0.23, we would need 565 pounds of gypsum per acre. This is basically treating the symptom and not the problem, but it is the cheapest and easiest fix for this problem. Repeating gypsum applications fall or spring will eventually start to bring the soils into a better balance.

What if the potassium in the paste test was 60 ppm instead of the 33.6 ppm, and the magnesium was 10 ppm and the high potassium was causing magnesium deficiencies to show up in the plants?. We could rebalance the potassium and magnesium ration to a more reasonable 2:1 (K:Mg) and raise the magnesium to 30 ppm and then push the calcium to 150 ppm in order to keep the calcium five times the magnesium. When doing this rebalancing. Gypsum and Epsom salts would be the most logical product choices. This can be done as long as the soluble salts don't get too high. If this is a possibility, do the rebalancing in the fall so the excess salts have a chance to leach out over the winter and spring prior to planting. Discovering this type of imbalance in the spring might mean foliar feeding some magnesium and calcium during the current growing season and then rebalancing in the fall.

Since the phosphorus is so high, it would be good to stay away from manure-based composts. Maintaining the organic matter could be accomplished with cover crops and mulching during the growing season with straw. Care must be taken when using high-lignin products like straw, since the carbon-to-nitrogen ratio is very wide and will tie up nitrogen when incorporated into the soil. Keeping the straw on the top of the soil as a mulch during the growing season will have a minimum effect on soil nitrogen. For those people fortunate enough to have soil organic matter levels above 10 percent, straw will help to tie up any excess nitrogen at the end of the growing season. For those people, with less than 10 percent soil organic matter, it would be good to add some nitrogen to aid in the decomposition of the straw.

The last thing in this example that should be addressed is the bicarbonates on the paste test. High bicarbonates can precipitate out calcium as the soils dry down and then the calcium carbonate can precipitate out the soluble phosphorus as a rock phosphate. The precipitation of calcium can lead to poor water infiltration and localized dry spots in the soil. The high bicarbonates are less important in this situation due to the high organic matter and the additional calcium that will be added for rebalancing. High bicarbonates generally come from the irrigation water. Drip irrigation tends to accumulate bicarbonates faster due to the lack of leaching. When watering, adding more water less often is the best habit to get into. Keeping the soil pH closer to 6.2 where irrigation is taking place will help to dissipate the bicarbonates. Buffering the water with acids at the well head will also reduce the bicarbonates. Humic or sulfuric acids can be used in this process, but a sample of the water along with the acid of choice should be sent to the lab and titrated to get the exact amounts needed.

Bicarbonates will tend to accumulate through the season as long as irrigation is taking place and the water source is coming from a deep well. Of course, with ample rainfall, this will help minimize the situation, but drought conditions will dramatically exacerbate the accumulation of bicarbonates. The soil solution tends to concentrate nutrients as soils dry out. This a good thing to a point, so let the plants experience a little stress and don't start the irrigation too soon, especially if the water source contains a lot of bicarbonates. The goal of using high-bicarbonate water is to use as little as possible until harvest. After harvest, applying a soil surfactant may help leach out excess bicarbonates during winter and spring rains.

The last example that we will discuss is one that we cannot balance on a standard soil report. This is a high-calcium soil that is located in the Southwest. There are many acres of naturally occurring high-calcium soils in the country. The coral-based soils in Florida and the high-calcium carbonate soils in Texas are two of the areas that I'm most familiar with. The opposite of the high-calcium soils would be the serpentine soils that are very high in magnesium and low in calcium.

Most of these soils are found in California and the eastern side of the Appalachian mountains. Both of these types of soils are deep and have huge reserves of their respective minerals. Therefore, changing their basic chemistry is virtually impossible. The best thing that we can do is to treat the symptom and not the problem. Example 8 is a high-calcium soil with extremely high reserves of calcium.

These types of high-calcium carbonate soils pose a real difficulty for someone trying to balance the nutrients, but it also poses a problem for the laboratory. The Mehlich extracting solution has a pH of 2.5 and will dissolve calcium carbonate, resulting in exaggerated calcium levels and exchange capacities. With enough calcium, the extracting solution will become neutralized and not extract phosphorus accurately. In these cases, I suggest having the lab extract the cations with ammonium acetate, which has a pH of 7.0. This greatly reduces the calcium and exchange capacity on the report, but it will still be high, and rightly so. Some people have gone as far as buffering the extracting solution to a pH of 8.4, which is the maximum pH that calcium carbonate will dissociate. The problem that I see with the 8.4 pH extracting solution is that it seems to underestimate the levels of magnesium and potassium.

Example 9 is an example of one high-calcareous soil using the three different extracting solutions. Regardless of what extracting solution you settle on, the numbers are relatively meaningless. If you use the Mehlich, the calcium is extremely overestimated as a colloidal nutrient. The ammonium acetate of 8.4 may actually do a better job of showing colloidal calcium, but with all the free carbonates waiting to dissolve, the odds are very good that if a calcium comes off the colloid, it will be replaced by another calcium cation. The base saturation percentages on all the extractions are very close. The colloidal cations will be a reflection of what is out in solution as well as what is potentially ready to go into solution. Therefore, I prefer to settle on the middle of the road in this situation and go with the regular ammonium acetate extraction for the standard test. The phosphorus values should be suspect for all calcareous soils. The Olsen extracting solution is quite commonly used in these high-calcium situations. The bottom line is that the paste test is the best method for determining what the plants see and what the best

Sample Location			12/13
Sample ID			South
Lab Number			32
Sample Depth in inches			6
Total Exchange Capacity (M.E.)			48.48
pH of Soil Sample			8.10
Organic Matter (percent)			5.39
ANIONS	SULFUR:	p.p.m.	21
	Mehlich III Phosphorous:	as (P_2O_5) lbs / acre	102
EXCHANGEABLE CATIONS	CALCIUM: lbs / acre	Desired Value	13186
		Value Found	16680
		Deficit	
	MAGNESIUM: lbs / acre	Desired Value	1396
		Value Found	864
		Deficit	-532
	POTASSIUM: lbs / acre	Desired Value	1512
		Value Found	1157
		Deficit	-355
	SODIUM:	lbs / acre	44
BASE SATURATION %	Calcium (60 to 70%)		86.02
	Magnesium (10 to 20%)		7.43
	Potassium (2 to 5%)		3.06
	Sodium (0.5 to 3%)		0.20
	Other Bases (Variable)		3.30
	Exchangeable Hydrogen (10 to 15%)		0.00
TRACE ELEMENTS	Boron (p.p.m.)		0.75
	Iron (p.p.m.)		24
	Manganese (p.p.m.)		96
	Copper (p.p.m.)		2.11
	Zinc (p.p.m.)		5.41
	Aluminum (p.p.m.)		209
OTHER			

Example 8: *Soil Report*

Sample Location			12/13
Sample ID			South
Lab Number			66531
Water Used			DI
pH			8.1
Solubale Salts		ppm	74
Chloride (Cl)		ppm	39
Bicabonate (HCO3)		ppm	102
ANIONS	SULFUR	ppm	3.27
	PHOSPHORUS	ppm	<0.03
SOLUBLE CATIONS	CALCIUM	ppm	16.07
		meq/l	0.80
	MAGNESIUM	ppm	1.8
		meq/l	0.15
	POTASSIUM:	ppm	3.69
		meq/l	0.10
	SODIUM	ppm	2.59
		meq/l	0.11
PERCENT	Calcium		69.15
	Magnesium		12.92
	Potassium		8.24
	Sodium		9.68
TRACE ELEMENTS	Boron (p.p.m.)		< 0.02
	Iron (p.p.m.)		0.51
	Manganese (p.p.m.)		< 0.02
	Copper (p.p.m.)		< 0.02
	Zinc (p.p.m.)		< 0.02
	Aluminum (p.p.m.)		0.76
OTHER			

Example 8A: *Saturated Paste Report*

approach is to balance the solution and not the soil.

The paste analysis of this calcareous soil (Example 8A) is quite low when compared to the ideal numbers. Even the calcium is low, but this is common on the high-pH soils. The solubility of calcium and magnesium carbonates are low at high pH levels; however, the feed rate of calcium may be adequate to supply all the calcium that is needed. The feed rate of magnesium will more than likely be inadequate, especially if the amount of magnesium on the standard test is low. In this example, both the magnesium and potassium values found on the standard test are quite high and adding anything more than enough to enhance the solubility would probably be wasteful. What really needs to be done in this situation is to treat the symptom by banding 200-400 pounds of KMag per acre to stimulate plant response, which will increase root exudates and consequently increase the microbial biomass. The increase in biological components will improve the availability of an already high level of potassium and magnesium. This is exactly what we did in this situation in Texas and saw a huge response in the crop production and quality. Cover crops should also be used to maintain a high level of biological activity.

Phosphorus in the form of mono-ammonium phosphate (11-52-0) was also banded in combination with a humic acid in order to improve the availability. Organic farmers are at a distinct disadvantage in these situations if manures or composted products are not available. Rock phosphate and bone meal are low-soluble products containing calcium, which only adds to the existing problem.

The trace elements look pretty normal on the standard test, but the paste test tells quite a different story. With the exception of iron, the rest of the traces are below the detection limit. Soil-applying traces will have a limited success, especially in a broadcast situation. Banding the traces along with humates has worked very well. I have done this with both liquid chelates and fulvic acid and also dry sulfated traces and dry humic acid. There are some companies who have impregnated trace elements on dry humic acid, which seems like a great idea.

Pin oaks trees in northwest Ohio generally start out very nice, but as they grow older and the roots grow more into the calcareous till, the

			Mehlich	AA	AA 8.2
Sample Location					
Sample ID			3		
Lab Number			4	5	6
Sample Depth in inches			6	6	6
Total Exchange Capacity (M.E.)			36.56	18.47	13.15
pH of Soil Sample			8.3	8.3	8.3
Organic Matter (percent)			3.73	3.73	3.73
ANIONS	SULFUR:	p.p.m.	11	11	11
	Mehlich III Phosphorous:	as (P_2O_5) lbs / acre	57	57	57
EXCHANGEABLE CATIONS	CALCIUM: lbs / acre	Desired Value	9943	5024	3576
		Value Found	13217	6627	4587
		Deficit			
	MAGNESIUM: lbs / acre	Desired Value	1052	532	378
		Value Found	493	267	245
		Deficit	-559	-265	-133
	POTASSIUM: lbs / acre	Desired Value	1140	576	410
		Value Found	184	121	142
		Deficit	-956	-455	-268
	SODIUM:	lbs / acre	43	29	33
BASE SATURATION %	Calcium (60 to 70%)		90.39	89.69	87.21
	Magnesium (10 to 20%)		5.62	6.02	7.76
	Potassium (2 to 5%)		0.65	0.84	1.38
	Sodium (0.5 to 3%)		0.25	0.35	0.55
	Other Bases (Variable)		3.10	3.10	3.10
	Exchangeable Hydrogen (10 to 15%)		0.00	0.00	0.00
TRACE ELEMENTS	Boron (p.p.m.)		0.74	0.74	0.74
	Iron (p.p.m.)		101	101	101
	Manganese (p.p.m.)		132	132	132
	Copper (p.p.m.)		8.61	8.61	8.61
	Zinc (p.p.m.)		9.02	9.02	9.02
	Aluminum (p.p.m.)		94	94	94
OTHER					

Example 9: *Soil Report*

trees begin to develop iron chlorosis on the leaves and sections of the trees will die back. Coring around the trees at the drip line and filling the cores with a combination of peat, dry humic acid, and iron sulfate has really helped these trees. It is impossible to change the entire soil, but a change in a small section of the soil seems to really help. Foliar feeding the traces is still the most efficient way to influence the minor element levels in the plant.

Chapter 14

Foliar Feeding

FOLIAR FEEDING NUTRIENTS IS A HIGHLY EFFECTIVE WAY TO CHANGE the nutrient balance within plants, but it is also the most expensive way. I prefer to do everything possible to improve soil balance and nutrient availability through the soil before turning to foliar feeding. Foliar applications of major and minor elements are often necessary during a plant's growth cycle in order to increase the level of a particular nutrient that is needed during a high-demand time for just a short period. These high-demand times may be any time during the growth cycle of a plant.

For example, tomatoes or peppers often need extra calcium during the first one or two flower sets in order to prevent blossom end rot. The soil levels may be good, but the plants are still growing vegetatively and the calcium sink is the foliage and not the flower. If the weather turns rainy with high humidity, this could reduce the movement of xylem-mobile nutrients like boron and calcium. There are many reasons to consider foliar feeding. Here is a list of some of them.

1. Excessive levels of interfering or competing nutrients in the soil. One example would be high levels of phosphorus, resulting in the tie-up of zinc. The preceding calcareous soil sample (8 and 8A) with excessive calcium could result in interference with potassium and magnesium. The high pH as a result of the excess calcium would also reduce the solubility of most of the trace elements.
2. Weather conditions can dramatically affect the uptake of nutrients. Excess moisture in the root zone will reduce uptake of nutrients or cause the loss of nutrients due to leaching. Denitrification of nitrogen under anaerobic conditions can happen quite rapidly in warm soils. High humidity will reduce evapotranspiration and

the uptake of xylem-mobile nutrients as mentioned earlier. Hail damage can greatly reduce a plant's leaf surface and expose the plant to diseases. If this occurs early in the growth cycle, I have seen a combination of nitrogen, sulfur, zinc, and fulvic acids help in the recovery and healing of the plant.

3. Sometimes all the ingredients for a great crop yield come together and the crop may need a little help to maintain internal nutrient levels in order to improve quality or prevent blossom drop or kernel abortion.

4. A tissue or sap analysis indicates a deficiency or an imbalance of nutrients in the plant. Remember, a plant grows to the least-available nutrient. Plants can look great vegetatively, but is there enough of the right nutrition to allow the plants to flower and max out reproduction? It is virtually impossible to foliar feed the major elements to any significant level, but sometimes it is not as much a matter of supplying a large portion of a major element as it is just keeping that nutrient above the threshold level before the plant begins to shut down. This may mean keeping the plant above the tipping point and running at maximum efficiency for just a few more days in order to boost yield and quality.

5. When using an herbicide like glyphosate, which is a metal chelating agent, it would be advisable to foliar feed a broad spectrum of trace elements, especially manganese. For the longest time, it was thought that glyphosate only chelated manganese, which was critical in the shikimate enzyme system for disease protection. Plants were not killed by the glyphosate, but a weakened immune system allowed diseases such as Rhizoctonia, Phytophthora, Pythium, and Fusarium to do the dirty work. It is now known that all the trace metals are at risk, and some major nutrients such as calcium and magnesium.

I'm sure there are more reasons, but these are a few to consider when growing any crop.

Plants receiving a foliar application will absorb the nutrients directly through the leaves. Plants can absorb nutrients through the stoma or directly through the epidermis. Absorption will be faster through the stoma than the epidermis. Dicots such as beans, tomatoes, and peppers have more stoma on the underside of the plant, whereas monocots such as corn, wheat, and onions have nearly equal numbers on top and bottom. It makes perfect sense when foliar feeding dicots to spray with higher pressure to create of fog and a turbulence in the canopy to get a good coating on both the top and bottom. The use of surfactants and penetrants can also help spread the products over the leaf surface and enhance penetration into the plant.

Plants under moisture stress are less likely to respond well to foliar feeding since the stoma are closed to preserve internal moisture. Remember that low potassium levels in the soil can lead to poor water utilization by the plant since potassium is critical for opening and closing the stoma.

Weather conditions can affect the success or failure of the foliar application.

1. The best time of day is generally in the morning after the dew has dried and before 11 a.m., or later in the day after 6 p.m. The stoma may be closed to prevent moisture loss during the heat of the day.
2. Spraying temperature is best between 60 and 85 degrees. It is advisable to hold off spraying if the previous nighttime temperature is significantly colder than 60 degrees.
3. Humidity is best when above 70 percent relative humidity. As the air temperature rises, more water in the atmosphere is needed to maintain the needed relative humidity. Therefore, in drier conditions, it is best to spray earlier in the day rather than later. The whole idea is to maintain enough humidity in the air, so as not to dry the foliar spray too quickly and leave the nutrients on the outside of the leaf.
4. Temperature/humidity index is often used as a guide for foliar feeding. An index of 140-160 is considered to be the ideal. This index is the sum of the temperature and the relative humidity.
5. Wind speeds under 5-10 mph is ideal. This will improve application and lower the evaporation rate of the spray.
6. Rainfall within 24-48 hours may reduce the effectiveness of the application.

The product list for foliar feeding is long and complicated. I primarily use the foliar feeding for minor elements; however, I believe there is a benefit for foliar feeding major elements during stress and peak demand times. There are many universities that claim there is a poor return on investment when it comes to foliar feeding trace elements. I have reviewed many of their studies, only to find that many studies had final yields less than 30 bushels per acres for soybeans. If you are getting soybean yields less than 45 and corn yields less than 140 bushels per acre, there is obviously something more yield restricting than minor elements. If you are going to use foliar feeding options to improve already good yields, then be willing to increase your major element nutrition program, especially nitrogen, to cover the potential yield increases. Plants grow to the least-available nutrient, so when fixing one, you should be automatically looking for the next one.

Sea-based foliar products are often used in organic production. Kelp is very popular and has many trace elements along with plant growth stimulants considered important in plant health and flowering. Fish emulsions are another popular organic-based product. Virtually all these types of products have recommended rates and should be followed according to the manufacturer's specification.

Rather than trying to correct a deficiency with a broad-spectrum product such as kelp, you can get more specific by foliar feeding sulfated trace metals such as zinc sulfate, copper sulfate, and manganese sulfate. Calcium, magnesium, and potassium sulfate may also be foliar fed, but the major problem lies with a foliage burn from the sulfate portion of the minerals. The literature suggests that sulfated forms of trace metals should be limited to 2-5 pounds dissolved in one hundred gallons of water. This level should prevent leaf scorch. Five pounds' total per hundred gallons of water for any combination of trace metals should be the upper limit. For example, if you wanted to foliar feed both zinc and manganese, you could add 2.5 pounds each or any combination of weights totaling 5 pounds or less per hundred gallons of water. It is always best to spray a test area before proceeding to the entire crop. When foliar feeding, it is best not to saturate the leaf to the point of run-off. Mist the plants. Foliar solutions that accumulates at the tip of the leaf may result in leaf burn, even at the recommended levels.

Chapter 15
Soil Nutrients vs. the Weather

THE WEATHER HAS A HUGE IMPACT ON NUTRIENT AVAILABILITY FOR A growing crop and, in the long run, the overall soil balance. Nutrients move through the soil at different rates. The rate of movement is dependent upon the nutrient itself and soil type — specifically the amount of sand, silt, clay and organic matter. Table 30 shows how plants pick up nutrients either through mass flow, diffusion, or direct root intercept. The nutrients that are picked up by mass flow — such as nitrogen, calcium, magnesium, sulfur, boron, and molybdenum — are the nutrients most easily moved through the soil profile. The nutrients picked up by diffusion, even though in solution, move very slowly in the soil, and the nutrients — primarily picked up by direct root intercept move hardly at all.

Organic matter is the stabilizing factor for all soils. It increases nutrient- and water-holding capacity for all soils, especially sand, and improves structure and granulation of silt and clay soils.

Within the different soil types, there is a wide range of available water for plant growth. The diagram below shows critical points when it comes to plant-available water and ultimately the nutrient transport for virtually all the nutrients except copper and half of the iron.

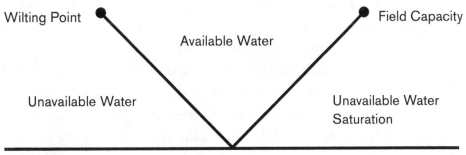

Table 30

The wilting point is the point at which plants can no longer take up water. Field capacity is the point at which all the gravitational water has drained out of the soil, leaving only the water held against gravity. The area between the wilting point and field capacity is water available for plant growth and nutrient absorption. For sandy soils, the distance between the wilting point and field capacity will be much narrower, while the clay and high organic matter soils will be much wider. It is important to note that compaction and poor soil structure in heavier soils will reduce available water. Each of the soils has their advantages and disadvantages.

Sandy soils, as well as very low organic soils, are very limited in their level of available water. This level of available water can effectively be reduced through over-fertilization. Nutrient levels in the soil solution will be concentrated as the soils begin to dry out. Too many nutrients in solution can create a salt effect and pull water back away from the plants. This fact is true for heavier soils as well. The concentration of nutrients as soils dry down will help the low-fertility soils during dry years when the plants are growing just above the wilting point, which results in an increase of the nutrient concentration. The odds that the weather pattern will help maintain this moisture level is rather unlikely. There are some things that can be done to help increase the soil's defense against drought.

1. Be sure there is enough calcium in the standard, as well as the paste, test. Calcium is critical for creating root mass, and the more roots, the more water that can be harvested. In heavier soils the calcium will not only help increase root mass but will also help flocculate clays, improving soil structure and increasing available water.
2. Keep the potassium at optimum levels. Potassium controls water utilization for the plant. Low levels of potassium will allow the stoma to remain open, wasting water.
3. Maintain an adequate phosphorus levels. As soils dry down and the salt concentration increases in the soil solution, plants will expend more energy to bring in water. Phosphorus is key in the ADP conversion to ATP, which is the energy used in the plant.

4. Establishing a plant canopy as quickly as possible will keep the direct sun off the soil, lowering the temperature and evaporation rate. An organic cover from previous crop residue or a cover crop will also improve water efficiency.

Soils with a higher clay and organic matter tend to have a better tolerance to drought, but excess water is as bad to the heavier soils as no water is to the lighter, sandier soils. What can be done if you find yourself in a protracted period of wet weather in the spring?

1. See number one above. Don't pull the trigger too quick when it comes to getting in the field. Compaction may come back to bite you later in the season when it typically dries out.
2. See number two above, but not for the same reason. Late planting effectively shortens the vegetative season since most plants flower based on light regime. In many plants, most of the potassium is taken up prior to flowering, so having good to excellent potassium levels helps minimize the reduced uptake period of a short vegetative period.
3. See number three above and if possible, when planting, band some soluble phosphorus in the row next to the plant. I would do this even if I had excellent levels. Saturated soils are anaerobic, which severely reduces biological activity. The soil biology is important for solubilizing phosphorus in the soil. Since their numbers will be reduced and the season is already short, banding some soluble phosphorus will really help get the plants off to a better start.
4. In certain situations, the use of a soil surfactant may help reduce excess water and allow planting sooner. This could be surface applied, in a starter or in a transplant solution.
5. Check for available nitrogen. Fall-applied nitrogen, even with a nitrogen stabilizer, can be at risk of being greatly reduced. Much of the normal soil nitrogen will be lost due to leaching or denitrification. Soils with a lot of residue will exhibit more nitrogen deficiency to a growing crop since the microbes will get first dibs for any available nitrogen in the soil. It would be a good idea to apply 15-20 pounds of nitrate and/or ammonium nitrogen to

the soil to help reduce the microbial competition. Even legumes could benefit from this practice since their nodule formation is a few weeks in the future after planting. Legumes could also benefit from a moly seed treatment or foliar application since moly is critical in nodulation, and being an anion, it is subject to leaching.

Droughts and excess rainfall is impossible to ever get completely prepared for; however, during my forty-three years of consulting, balanced soils have come through stress situations much better than unbalanced ones.

Chapter 16

Modified Growing Media

DEVELOPING A GROWING MEDIA IS MUCH MORE DIFFICULT THAN ONE might think. The mix not only has to provide an adequate growing medium for root development, but also an adequate water and nutrient holding capacity. Here are some suggestions to consider when developing a Modified Growing Media, or MGM. These are based on the test result that I have seen coming through the lab and conversations that I have had with growers.

Mix only the primary ingredients, like peat, perlite, silica sand, washed coir fiber, etc. This mix must be heavy enough to support the plants you intend to grow as well as maintain adequate moisture for plant growth. This mix must also have good enough drainage to allow for the leaching of soluble salts. High organic mixes tend become hydrophobic if they are allowed to dry out. High bicarbonate irrigation water will exacerbate this problem. This issue will make watering the mix more difficult and often leads to inconsistent moisture levels within beds or containers. Therefore, creating a mix with good water infiltration is very important. The use of a good soil surfactant might become necessary should the surface become hydrophobic.

Send the sample of primary ingredients to the laboratory for a standard soil analysis and paste analysis. The standard test will help us develop the mineral balance initially. We won't use the paste test too much till the second round of sampling, but I would still run a paste test on the base mix to see what it might be contributing to the salts, chlorides, and overall soluble nutrients. If this sample is run through Logan Labs and you inform them on your worksheet that this is a modified growing media, the standard test will also include a bulk density reading. You must have a bulk density number to estimate nutrient supplementation. The bulk density number will indicate how your mix compares to that of a normal field soil. Generally, the bulk density of MGM will be less than 30 percent of normal soil.

Let's assume the bulk density of your mix is 0.25, which means the weight of the mix is 25 percent of a normal soil. For formulating a modified growing media, you should start by multiplying whatever amounts you would use to balance a normal soil by the bulk density, in this case 0.25. For example, let's assume that on the standard test of the media that you got back from the lab that it had a calcium deficit, you would normally use 2,000 pounds of high-calcium lime to correct the pH and calcium level. You would then multiply the 2,000 pounds by 0.25, ending up with 500 pounds of high-calcium lime to correct an acre furrow slice. An acre furrow slice is 6 inches by 43,560 square feet, or 21,780 cubic feet (43,560 divided by 2). If you were making a 400-cubic-foot mix, how much calcium would you need? 400 divided by 21,780 means that your 400-cubic-foot mix is just 1.8 percent of the acre furrow slice. Multiplying 1.8 percent times 500 pounds of lime means that you would need to add 9 pounds of high-calcium lime to your 400-cubic-foot mix.

The fineness of the lime will have a tremendous effect on the availability of the lime. These calculations are meant to balance the standard test, but ultimately the amount of calcium moving into solution will decide whether or not more or less lime will be used. This method could be applied to all the cations. Organic sources of phosphorus that are low in solubility can be calculated the same way, but the bulk density number needs to be doubled at least. In this case, we would use 0.5 for the media density instead of 0.25. Keep in mind this is a starting point; depending upon the solubility of the products chosen, you may have to use more or less. Growing conditions, water quality, and varieties will all play a role as to the ultimate desired levels you will need to settle on. This formula seems to work well in keeping the mixes from becoming too rich in nutrients and high in soluble salts. Ideally I would like to keep the soluble salts down to 300–400ppm for direct seeding and 400–500ppm for transplanting small plants. Towards the end of the grow cycle, mature plants could easily handle 700-800ppm of soluble salts. Soluble salts does not only refer to things like sodium and chlorides; it also refers to those nutrients necessary for growth, such as potassium, calcium, magnesium, sulfur etc. It is quite possible to have too low of a level of soluble salts/nutrients in solution towards the end of the grow cycle which will result in limiting growth and quality.

The Mehlich 3 extracting solution used by the lab for the standard soil analysis was really not intended to be used for soil-less media. These artificial mixes will have virtually no colloidal characteristics — just nutrients floating in organic and drainage-enhancing substrates. The exchange capacity is merely the summation of cations found floating in an organic medium and by no means an actual colloidal exchange capacity.

The reason for this early round of sampling is two-fold. First, it is important to see just what the base mix is contributing to the overall nutrition and pH level. Secondly, the bulk density number will give you a guideline to how much nutrition can be added without creating a soluble salts problem from over-fertilization.

The pH of the final mix should end up around 6.5 and the soluble salts less than 500 ppm for starting small plants. This level can increase proportionally to the size of plants being transplanted. Hold the soluble salts level down if the irrigation water is on the high side in soluble salts.

Balance the calcium-to-magnesium ratio in the range of 5-6 to 1 and a base saturation ratio on the standard test of 60-65 percent calcium and magnesium 12-15 percent.

Balance the magnesium-to-potassium ratio in the range of 1 to 1.2-1.5 and the base saturation ratio on the standard test of 12-15 percent for magnesium and potassium 4-6 percent. Potassium is poorly held by the organic constituents of the media mix and moves into solution very easily. Potassium excess in solution is one of the primary reasons that I see high salt levels in the paste extract.

Add the appropriate level of compost, worm castings, and any other nutritional products, keeping in mind the bulk density of the mix. Of course, it is important to know the analysis of whatever you are adding to the mix, not just average values from a book, but actual analysis —

especially for things like composts and worm castings. Figure 1 and 2 is a soil and a paste analysis of an actual compost sample. I don't recommend running a standard soil analysis on compost, fish meal, or any straight additive product. The test was not designed for those types of samples. I might make an exception for worm castings. Looking at the two tests, and especially the paste analysis, you can see how a little bit of this compost could elevate the pH, bicarbonates, soluble salts, and potassium very rapidly. A 5 percent inclusion rate of this compost might be the upper limit that you might add, depending upon the base mix.

The length of the grow cycle will depend on the degree of solubility needed in your product choice. Using things like rock phosphate or green sand, which are very low solubility products, may not feed the crop fast enough, and if the MGM is a one-crop product, it will more than likely be left in the mix when you are done. If the media is going to be used for more than one crop, 10-15 percent of the low-soluble nutrients might be a good choice.

Another issue quite common with modified growing media is getting a thorough and complete mix of all the ingredients. The base mix materials and meals should mix fairly well, but minerals can be difficult. If you cannot get a good mix, it will be difficult to get a good sample for analysis. I have talked to growers who had pots that were growing very well, while others were struggling even though the media was from the same batch. The base mix materials and meals may weigh 20-30 pounds per cubic foot whereas the minerals could easily weigh 60 pounds per cubic foot. There will always be sifting and settling of the heavy minerals, especially when the mix is drier. Wetting the base mix will help to stick the minerals to the organic products. Mixing smaller batches — as opposed to one great big mix — will help if you are not using some sort of horizontal mixer. Micro-ingredients such as the trace elements and microbes would be best dissolved in water and sprayed on during the mixing process.

Once the MGM is thoroughly mixed, it is time to resample for another standard test and paste test. Figure 2 indicates the starting points that I would like to see in the paste analysis. This is only a starting point, and tissue analysis should be used to refine the numbers. Yields, geo-

Standard Soil Analysis of a Compost Sample

Sample Location			Compost
Sample ID			
Lab Number			36
Sample Depth in inches			6
Total Exchange Capacity (M.E.)			26.85
pH of Soil Sample			9.1
Organic Matter (percent)			74.27
ANIONS	SULFUR:	p.p.m.	177
	Mehlich III Phosphorous:	as (P_2O_5) lbs / acre	5469
EXCHANGEABLE CATIONS	CALCIUM: lbs / acre	Desired Value	7302
		Value Found	3765
		Deficit	-3537
	MAGNESIUM: lbs / acre	Desired Value	773
		Value Found	1668
		Deficit	
	POTASSIUM: lbs / acre	Desired Value	837
		Value Found	6296
		Deficit	
	SODIUM:	lbs / acre	826
BASE SATURATION %	Calcium (60 to 70%)		35.06
	Magnesium (10 to 20%)		25.89
	Potassium (2 to 5%)		30.06
	Sodium (0.5 to 3%)		6.69
	Other Bases (Variable)		2.30
	Exchangeable Hydrogen (10 to 15%)		0.00
TRACE ELEMENTS	Boron (p.p.m.)		1.58
	Iron (p.p.m.)		76
	Manganese (p.p.m.)		31
	Copper (p.p.m.)		0.61
	Zinc (p.p.m.)		18.71
	Aluminum (p.p.m.)		21
OTHER	Media Density g/cm3		0.57

Figure 1: *Soil Report*

Saturated Paste of a Compost Sample

Sample Location			Compost
Sample ID			
Lab Number			127499
Water Used			DI
pH			9.1
Solubale Salts		ppm	2,557
Chloride (Cl)		ppm	900
Bicabonate (HCO3)		ppm	872
ANIONS	SULFUR	ppm	45.61
	PHOSPHORUS	ppm	29.69
SOLUBLE CATIONS	CALCIUM	ppm	73.49
		meq/l	3.67
	MAGNESIUM	ppm	53.11
		meq/l	4.43
	POTASSIUM:	ppm	1,065.00
		meq/l	27.66
	SODIUM	ppm	96.59
		meq/l	4.20
PERCENT	Calcium		9.19
	Magnesium		11.08
	Potassium		69.22
	Sodium		10.51
TRACE ELEMENTS	Boron (p.p.m.)		0.47
	Iron (p.p.m.)		3.83
	Manganese (p.p.m.)		0.6
	Copper (p.p.m.)		0.72
	Zinc (p.p.m.)		2.18
	Aluminum (p.p.m.)		0.37
OTHER			

Figure 2: *Saturated Paste Report*

graphical locations, water quality and plant varieties could have a big difference on the needs of the various crops and the balance needed in the growing media. One-mix adjusting example that comes to mind is the quality of the water that will be used for irrigation. If the irrigation water is coming from a well that is high pH due to calcium, magnesium, and bicarbonates in the water, it might be best to try and start the pH of your mix off at around 5.8. It is easy to have enough a calcium and magnesium in solution at this lower pH if you shift away from limes in the initial blend and use more gypsum and Epsom salts. The traces will be very available along with most everything else. As you start watering with the high pH water, the bicarbonates will be neutralized and the young, less tolerant plants will get off to a good start.

It is the cheapest and easiest to balance calcium and magnesium by using as much high-calcium or dolomite agricultural lime as possible; just don't raise the pH over 6.5. If you need more soluble calcium, gypsum fits the bill nicely. More soluble magnesium can be supplied with Epsom salts or KMag unless the potassium is already high. Borax and the sulfated form of traces would be the cheapest. Chelated traces could also be used, but they would fit if the mix is to be totally organic.

Organic growers need to try to get their available phosphorus through micronized products, meals, or compost.

Guidelines for Modified Growing Media

	Desired Levels Pounds per 500ft³ or 1000ft² @ 6"
Soil pH	6.5
Sulfur ppm	20-25
Mehlich III Phosphorus (P_2O_5) lbs.	12-14
Calcium lbs	50-80
Magnesium lbs	9-12
Potassium lbs	6-10
Sodium lbs	1-3
Base Saturation %	
Calcium %	60-65
Magnesium %	15-20
Potassium %	4-6
Sodium %	<2
Traces	
Boron ppm	0.7-1.0
Iron ppm	>50
Manganese ppm	25-30
Copper ppm	2.5-3.5
Zinc ppm	6-8
Aluminum ppm	<600

Figure 3

Paste Guidelines for Modified Growing Media

Soil pH	6.2-6.5
Soluble Salts ppm	<500
Anions	
Sulfur ppm	1-3
Chlorides ppm	<60
Bicarbonates ppm	<90
Phosphorus ppm	1.5-3.0
Cations	
Calcium ppm	50-70
Magnesium ppm	10-14
Potassium ppm	20-25
Sodium ppm	<6
Solution %	
Calcium %	62
Magnesium %	20
Potassium %	12-15
Sodium %	<5
Trace Elements	
Boron ppm	0.08-0.10
Iron ppm	0.5-1.5
Manganese ppm	0.08-0.12
Copper ppm	0.06-0.10
Zinc ppm	0.10-0.15
Aluminum ppm	<1.5

Figure 4

Figures 5-9 is a modified growing media that was sent to me that was having some problems. The plants were performing poorly and exhibiting burning on the edges of the leaves. Figure 5 is the media done as a standard soil analysis and the cations and phosphorus reported as pounds per acre. Figure 6 is the same media, only reported in pounds per 1000 ft². Figure 8 is a standard soil analysis after the media was flushed with water. It too is reported in pounds per thousand square feet. The standard analysis assumes a sampling depth of 6 inches. These reports could also be considered as pounds per 500 cubic feet, since it is the same as pounds per thousand square feet sampled at 6 inches. Pounds per cubic foot might benefit people better who are growing in pots or raised beds.

The first set of standard and paste analysis is when the sample arrived at the lab and it was determined that there was a salts problem. Many of the samples I get are having problems related to over-fertilizing and high salts. Salts issues show up as a burning or necrosis on the edges of the leaves. During hot days, the plants may start to wilt earlier than normal even if you watered in the morning. Ideally the media mix should be below 400 ppm for seed germination, but larger plants seem to tolerate salts up to the 700-800 ppm range. Once soluble salts reach 1,000 ppm, large plants will begin to suffer. Plants expend more energy trying to grow and bring in water into the roots in a high-salts environment. If this is a routine problem due to marginal water quality, increasing the soil phosphorus levels and making sure those nutrients like potassium, manganese, and silicon, which improve water efficiency, are at optimum levels, but not in excess. This would allow for less watering, reducing the buildup of bicarbonates and salts.

The standard soil report has a number of indicators that soluble salts could be a problem. The organic matter and the media density is a very good indicator that this is a MGM. The high sulfur, potassium, and sodium all give a good possibility of a soluble salts problem, which is confirmed in Figure 7 on the paste test. The high organic matter seems to offer very little buffering capacity in these types of mixes. The potassium and sodium are not held by the organic matter and tend to stay in solution. Potassium in these mixes on the standard soil report should probably to be in the 300-400 pounds per acre range and sodium less than 100 pounds per acre to stay out of the excessive side in the paste analysis. Nearly 30 percent of these two cations will go directly into solution.

MGM Standard Soil Report, Pounds/Acre

	Sample Location		RM-1
	Sample ID		
	Lab Number		11
	Sample Depth in inches		6
	Total Exchange Capacity (M.E.)		18.26
	pH of Soil Sample		6.5
	Organic Matter (percent)		32.66
ANIONS	SULFUR:	p.p.m.	499
	Mehlich III Phosphorous:	as (P_2O_5) lbs / acre	943
EXCHANGEABLE CATIONS	CALCIUM: lbs / acre	Desired Value Value Found Deficit	4966 4764 -202
	MAGNESIUM: lbs / acre	Desired Value Value Found Deficit	525 584
	POTASSIUM: lbs / acre	Desired Value Value Found Deficit	569 802
	SODIUM:	lbs / acre	286
BASE SATURATION %	Calcium (60 to 70%)		65.23
	Magnesium (10 to 20%)		13.33
	Potassium (2 to 5%)		5.63
	Sodium (0.5 to 3%)		3.41
	Other Bases (Variable)		4.90
	Exchangeable Hydrogen (10 to 15%)		7.50
TRACE ELEMENTS	Boron (p.p.m.)		0.72
	Iron (p.p.m.)		78
	Manganese (p.p.m.)		10
	Copper (p.p.m.)		0.42
	Zinc (p.p.m.)		4.36
	Aluminum (p.p.m.)		69
OTHER	Media Density g/cm3		0.22

Figure 5: *Soil Report*

MGM Standard Test Before Flushing, Pounds/1,000 ft²

Sample Location			RM-1
Sample ID			
Lab Number			11
Sample Depth in inches			6
Total Exchange Capacity (M.E.)			18.26
pH of Soil Sample			6.5
Organic Matter (percent)			32.66
ANIONS	SULFUR:	p.p.m.	499
	Mehlich III Phosphorous:	as (P_2O_5) lb / 1000 sq ft	22
EXCHANGEABLE CATIONS	CALCIUM: lb / 1000 sq ft	Value Found	109
	MAGNESIUM: lb / 1000 sq ft	Value Found	13
	POTASSIUM: lb / 1000 sq ft	Value Found	18
	SODIUM:	lb / 1000 sq ft	7
BASE SATURATION %	Calcium (60 to 70%)		65.23
	Magnesium (10 to 20%)		13.33
	Potassium (2 to 5%)		5.63
	Sodium (0.5 to 3%)		3.41
	Other Bases (Variable)		4.90
	Exchangeable Hydrogen (10 to 15%)		7.50
TRACE ELEMENTS	Boron (p.p.m.)		0.72
	Iron (p.p.m.)		78
	Manganese (p.p.m.)		10
	Copper (p.p.m.)		0.42
	Zinc (p.p.m.)		4.36
	Aluminum (p.p.m.)		69
OTHER			

Figure 6: *Soil Report*

MGM Paste Test before Flushing

Sample Location			RM-1
Sample ID			
Lab Number			127164
Water Used			DI
pH			6.5
Solubale Salts		ppm	1,201
Chloride (Cl)		ppm	50
Bicabonate (HCO3)		ppm	146
ANIONS	SULFUR	ppm	88.39
	PHOSPHORUS	ppm	5.35
SOLUBLE CATIONS	CALCIUM	ppm	183.90
		meq/l	9.20
	MAGNESIUM	ppm	55.58
		meq/l	4.63
	POTASSIUM:	ppm	124.80
		meq/l	3.24
	SODIUM	ppm	39.42
		meq/l	1.71
PERCENT	Calcium		48.96
	Magnesium		24.66
	Potassium		17.26
	Sodium		9.13
TRACE ELEMENTS	Boron (p.p.m.)		0.13
	Iron (p.p.m.)		0.11
	Manganese (p.p.m.)		0.18
	Copper (p.p.m.)		< 0.02
	Zinc (p.p.m.)		0.02
	Aluminum (p.p.m.)		0.12
OTHER			

Figure 7: *Saturated Paste Report*

MGM Soil Test after Flushing, Pounds/1,000 ft²

Sample Location		RM-1
Sample ID		
Lab Number		23
Sample Depth in inches		6
Total Exchange Capacity (M.E.)		7.80
pH of Soil Sample		7.1
Organic Matter (percent)		41.03
ANIONS	SULFUR: p.p.m.	36
	Mehlich III Phosphorous: as (P_2O_5) lb / 1000 sq ft	18
EXCHANGEABLE CATIONS	CALCIUM: lb / 1000 sq ft Value Found	49
	MAGNESIUM: lb / 1000 sq ft Value Found	8
	POTASSIUM: lb / 1000 sq ft Value Found	6
	SODIUM: lb / 1000 sq ft	3
BASE SATURATION %	Calcium (60 to 70%)	69.00
	Magnesium (10 to 20%)	18.17
	Potassium (2 to 5%)	4.29
	Sodium (0.5 to 3%)	4.21
	Other Bases (Variable)	4.30
	Exchangeable Hydrogen (10 to 15%)	0.00
TRACE ELEMENTS	Boron (p.p.m.)	0.48
	Iron (p.p.m.)	55
	Manganese (p.p.m.)	6
	Copper (p.p.m.)	0.37
	Zinc (p.p.m.)	3.48
	Aluminum (p.p.m.)	72
OTHER		

Figure 8: *Soil Report*

MGM Paste Test after Flushing

Sample Location			RM-1
Sample ID			
Lab Number			128232
Water Used			DI
pH			7.1
Solubale Salts		ppm	255
Chloride (Cl)		ppm	20
Bicabonate (HCO3)		ppm	138
ANIONS	SULFUR	ppm	11.53
	PHOSPHORUS	ppm	9.1
SOLUBLE CATIONS	CALCIUM	ppm	28.47
		meq/l	1.42
	MAGNESIUM	ppm	9.20
		meq/l	0.77
	POTASSIUM:	ppm	38.68
		meq/l	1.00
	SODIUM	ppm	18.50
		meq/l	0.80
PERCENT	Calcium		35.59
	Magnesium		19.18
	Potassium		25.12
	Sodium		20.11
TRACE ELEMENTS	Boron (p.p.m.)		0.05
	Iron (p.p.m.)		0.25
	Manganese (p.p.m.)		0.02
	Copper (p.p.m.)		0.03
	Zinc (p.p.m.)		0.04
	Aluminum (p.p.m.)		0.12
OTHER			

Figure 9: *Saturated Paste Report*

As you can see from Figures 8 and 9, it is possible to lower the salts levels through flushing. When this is done, you will be flushing out all cations and not just sodium or potassium. The paste test in Figure 9 shows that we actually took the flushing a little too far since the calcium and magnesium numbers are a little low and should be readjusted. The phosphorus did not change; in fact, it went up, but this could be a just a sampling issue. This is why, when building a media mix, it needs to have good external drainage. Flushing a mix will take lots of water and using a soil surfactant will aid in process. I have suggested to growers to use 2-3 tablespoons of Dawn dishwashing soap in a two-gallon sprayer and spray the beds prior to flushing. This will act as a surfactant, breaking down the surface tension of the water and allowing the water to better penetrate through the media.

How do you know when to stop flushing? By collecting the water as it leaves the bed, you can analyze using an EC meter. When the meter reads around 0.6 mmhos/cm, which is equivalent to around 400 ppm soluble salts. This should be good enough unless your water quality is rather marginal; then you might want to take it down a little further. If you are transplanting larger, more established plants, the salt concentration could be higher, but you must keep in mind the length of time remaining in the grow cycle and the amount of water that will be needed. Remember, you can only flush the soluble salts down in the media to the level that is in your irrigation water. So, if your irrigation water is 500 ppm soluble salts, that is the lowest level you will be able to get on the mix.

Many people that I have talked to who are growing in beds raise off the ground are starting off without any regard for drainage. A lot of people have indicated that they have brought into their hoop house a mixture of compost and organic products and built their bed on the existing ground. Anytime a coarser material is laid over a finer base material, the percolating water through the media will initially be pulled into the base mix, but due to the fineness of the base mix, the water will tend to back up and puddle. In this situation, you are left with only one option and that is to water just enough to dampen the media. Even with fairly good water, salts will begin to build up. Assuming your water is 300 ppm soluble salts, after four irrigation cycles with no flush, the soluble salts would be nearly 1,200 ppm.

Building the beds over coarser material such as stone or sand may be a little better, but the tendency is for the finer media material to become saturated before dripping into the coarser material. This results in a partial saturation of the media and anaerobic conditions for the plant roots. In order to prevent this again the only choice is limited watering cycles.

The ideal solution is to put in drainage before the beds are constructed. Assuming that you are going to build on an existing soil and the site is very level, start by sloping the base soil to the center of each bed and sloping the length of the bed to a collection point. On a short hoop house maybe grade to one end. On longer houses, grade to central points to reduce the amount of soil needed to be excavated. Cross tile will need to be laid in and taken outside the hoop house to a collection point and pumped away. Lay in perforated tile in the center of the bed and build the beds over the tile (see Figures 10 and 11).

End View for Drainage

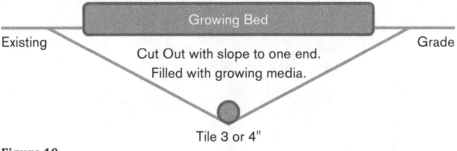

Figure 10

Side View for drainage

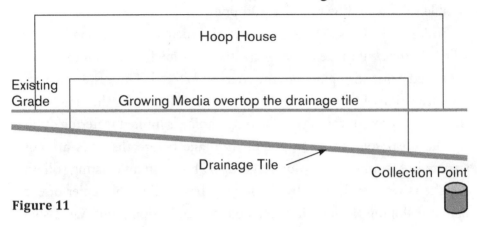

Figure 11

With tile in place you can constantly add enough water to obtain drainage and help prevent the buildup of salts. With this type of system, the media could be reused many times over, rather than grow for one cycle and then throw it away.

Chapter 17

Water Analysis

IN THIS SECTION I WANT TO DISCUSS WATER AND ITS EFFECT ON SOILS and media mixes. The water analysis is in order of best to worst, but I need to make clear there is water much worse than these last two examples. First let's look at the standard guidelines for water. The following information is based on references from *Water Quality for Irrigation* by L.K. Stromberg; 1975 and 1980 and *Water Quality for Agriculture* F.A.O. 1976.

Electrical Conductivity (EC) mmhos/cm

Level of Concern	Description
Below 0.75	Low salinity risk. Good for most crops
0.75-1.5	Medium salinity risk. Need moderately salt-tolerant crops.
1.5-3.0	High salinity risk. Good only for salt-tolerant crops.
Greater than 3.0	Very high risk. Not recommended

Table 32

EC times 640 equals total dissolved solids TDSin ppm. This is often referred to as soluble salts.

Sodium Adsorption Ratio (SAR): This an estimation of the amount of sodium absorbed on the exchange sites after extended use of the irrigation water.

Adjusted Sodium Adsorption Ratio: This is a better sodium adsorption indicator when bicarbonates and carbonates are present in the water. This is an attempt to adjust for the precipitation of calcium and magnesium when bicarbonates and carbonates are high. Permeability problems are more probable at a given SAR_{adj} with waters of low salinity rather than high salinity. Sodium is such a large cation that it holds clays apart, reducing flocculation and resulting in a permeability problem. This destruction of the soil structure results in poor water infiltration and anaerobic conditions. This situation can harm plants before sodium becomes toxic to the plant. The adjusted sodium adsorption ratio is calculated using a pHc, which is based on the total calcium, magnesium, bicarbonates, and carbonates. The pHc is a good indicator to whether lime will be deposited or dissolved in the soil. A pHc value above 8.4 will tend to dissolve lime and a value below 8.4 will tend to precipitate lime as the water moves through the soil.

SAR_{adj}	SAR	Effect
Below 6	Below 10	No permeability problem
6-9	10-15	Possible permeability problems
Above 9	Above 15	Definite permeability problems

Table 33

Irrigation water can pose structural and permeability problems in soil, along with a host of nutritional issues. Bicarbonates can tie up calcium and iron, creating nutritional deficiencies, but high levels of chlorides and boron can create toxicity issues very quickly. The following two tables show the acceptable levels for irrigation water.

Chloride meq/L

Below 2	Good for all crops
2-10	Possible leaf burn on chloride sensitive plants
Above 10	Unacceptable as irrigation water

Table 34

Many fruit trees and landscape woody ornamentals are very chloride sensitive.

Boron ppm

Below 0.5	Good for all crops
0.5-1.0	Sensitive may show injury but yields may not be effected.
1.0-2.0	Satisfactory for only semi-tolerant crops.
2.0-4.0	Satisfactory for only tolerant crops.
Above 4.0	Unacceptable for continual use.

Table 35

Irrigation water can have a huge impact on soil structure and nutrient availability. Just looking at water quality on a water analysis tells only part of the story. Parameters like SAR assume some leaching and a continual movement of things like sodium out of the root zone. With overhead irrigation that is certainly possible, but the conservation of our water resources has led to more drip irrigation and virtually no leaching, except through rain events. Even the leaching of excessive nutrients through rain events has been curtailed by the use of plastic mulches. When irrigating under these circumstances, the soil solution begins to look a lot like the irrigation water after just four or five cycles. I like to look at the water on a cation balance approach. By summing the milliequivalents of the cations and dividing each of the cations by the total, you can see what direction your cation balance will be heading as a result of the irrigation water. Table 36 shows the cation balances of the water samples 2 & 3.

	Water Sample 2		Water Sample 3	
Cation	Milli-equivalent	% Base Saturation	Milli-equivalent	% Base Saturation
Calcium	2.64	83.8	1.21	52.9
Magnesium	0.28	8.9	0.11	4.8
Potassium	0.02	0.6	0.06	2.6
Sodium	0.21	6.7	0.91	39.7
Total	3.15	100	2.29	100

Table 36

Assuming that a soil was balanced prior to irrigation, then under an irrigation regiment design to leach, you can see that Sample 2 will have the tendency to push magnesium and potassium out of the solution and replace it with a lot of calcium and a small amount of sodium. Even under drip irrigation, your balance will head the same direction but will quickly begin to precipitate out calcium and reduce phosphorus and trace element availability — especially manganese and iron. Soil structure won't be impaired and permeability will be sustained. Just the nutrient balance will suffer. The soluble salts will accumulate to a critical level after five to six irrigation cycles, assuming no leaching.

There are some things that could be done to improve the soil balance given enough lead time prior to irrigation. First allow the soil pH to drop to around 5.8 or 6.0. This would help to buffer the high-pH water and neutralize some of the initial bicarbonates. Fertilizing with KMag prior to planting (or better yet, just prior to starting the irrigation) would help to ramp up the magnesium and potassium and help offset the high-calcium water. Also, emitters may plug due to the potential for calcium precipitation. Acidifying the water at the wellhead with sulfuric acid or organic acids would help reduce this problem.

Water Sample 3 is different in its cation balance. This water has a lower salt concentration, but the higher sodium balance and concentration will eventually have an impact on the soil structure, especially with drip irrigation. The clays will come apart and internal drainage will be slowed down, possibly leading to anaerobic conditions. The addition of gypsum to this water would improve the balance and reduce the effect of sodium on deflocculating clays. The pHc is below 8.4 indicating the tendency to precipitate calcium, and the gypsum would help prevent this. The adjusted SAR indicates no potential for permeability problems on the short-term basis, especially if some leaching could occur. If soluble salts at 1,000 ppm is the magic number, then this water in Sample 3 would not hit that number until seven or eight non-leaching irrigation cycles, whereas Sample 2 would hit that in five or six cycles. This might require 8-10 weeks of irrigation, and the crop could be close to maturity by that time. The older and bigger the crop gets, the better it can withstand a higher salt concentration. This water is really pretty good when compared to some really bad water like water Sample 4.

Growing plants with Sample 4 would require growing salt-tolerant plants and/or a lot of water treatment before this water could be used on a regular basis.

In general, the deeper the wells, the higher the potential for water issues to show up. Many of the deep wells in the West are plagued with high salt concentrations.

The rain water in Example 1 would be the ideal water for irrigation, but it does have one interesting point to consider. Notice the pHc is 9.3, which is above the 8.4 break point. This means that as water percolates through the soil it is likely to strip calcium from the soil. This will be the most soluble of the calcium in the soil profile. This is why many of the soils that I test are low in soluble calcium. You can see the results of this calcium stripping for yourself if you test the soil that has been dipped from a ditch which is being cleaned. The soil pH will be around 7.5 and the calcium more than likely be above 80 percent base saturation.

Water Analysis Report

Job Name	SoilTech	
Contact	Bill McKibben	
Rep		
Submitted By	Bill McKibben	

Company	SoilTech	
Sample ID	304521	
Lab Number	12536	
Run Date	11/9/2017	

Sample Location: Rain
Sample Name:

Notes

pH		6.6
Hardness	ppm	4.9
Hardness Grains	/gal	0.28
Conductivity	mmhos/cm	0.02
Sodium Adsorption Ratio		0.41

		ppm	meq/L	lbs/A in
Calcium	Ca	1.3	0.06	0.29
Magnesium	Mg	0.4	0.03	0.09
Potassium	K	< 0.3	0.00	0.04
Sodium	Na	2.1	0.09	0.48
Iron	Fe	< 0.1		0.01

		meq/L	lbs/A in
Total Alkalinity	3.0		0.68
Carbonate	0.0	0.00	0.00
Bicarbonate	6.0	0.10	1.36
Chloride	2.0	0.06	0.45
Sulfate	2.0	0.04	0.45
Salt Concentration	12.2		2.76
Boron	0.02		
Cation/Anion Ratio		0.99	
pHc	9.31		
Adj. SAR	0.04		

Water 1

Water Analysis Report

Job Name	Palmetto Harmony	Company	Palmetto Harmony
Contact		Sample ID	337964
Rep		Lab Number	14205
Submitted By		Run Date	1/21/2019

Sample Location	Well		Notes
Sample Name			
pH		7.7	
Hardness	ppm	145.5	
Hardness Grains	/gal	8.51	
Conductivity	mmhos/cm	0.29	
Sodium Adsorption Ratio		0.18	

		ppm	meq/L	lbs/A in
Calcium	Ca	52.8	2.64	11.99
Magnesium	Mg	3.3	0.28	0.76
Potassium	K	0.7	0.02	0.15
Sodium	Na	4.9	0.21	1.12
Iron	Fe	0.1		0.02

	ppm	meq/L	lbs/A in
Total Alkalinity	145.0		32.95
Carbonate	0.0	0.00	0.00
Bicarbonate	177.0	2.90	40.23
Chloride	8.0	0.23	1.82
Sulfate	15.3	0.32	3.49
Salt Concentration	185.6		42.18
Boron	< 0.02		
Cation/Anion Ratio		0.91	

Nitrate ppm	0.2
P ppm	0.013
pHc	7.68
Adj. SAR	0.3

Water 2

Water Analysis Report

Job Name	Igor Rakuz	
Contact		
Rep		
Submitted By		

Company	Igor Rakuz	
Sample ID	337096	
Lab Number	14118	
Run Date	1/7/2019	

Sample Location   Water   Notes
Sample Name

pH		6.7
Hardness	ppm	65.7
Hardness Grains	/gal	3.84
Conductivity	mmhos/cm	0.22
Sodium Adsorption Ratio		1.12

		ppm	meq/L	lbs/A in
Calcium	Ca	24.1	1.21	5.49
Magnesium	Mg	1.3	0.11	0.30
Potassium	K	2.2	0.06	0.50
Sodium	Na	20.9	0.91	4.75
Iron	Fe	< 0.1		0.01

		meq/L	lbs/A in
Total Alkalinity	79.0		17.95
Carbonate	0.0	0.00	0.00
Bicarbonate	96.0	1.57	21.82
Chloride	9.0	0.26	2.05
Sulfate	18.1	0.38	4.12
Salt Concentration	140.8		32.00
Boron	< 0.02		
Cation/Anion Ratio		1.03	

Nitrate ppm	0.2
P ppm	0.015
pHc	8.15
Adj. SAR	1.4

Water 3

Water Analysis Report

Job Name	Soil Tech		Company	Soil Tech
Contact			Sample ID	341492
Rep			Lab Number	14408
Submitted By			Run Date	3/19/2019

Sample Location	Water		Notes
Sample Name			

pH			7.3
Hardness	ppm		790.3
Hardness Grains	/gal		46.22
Conductivity	mmhos/cm		2.70
Sodium Adsorption Ratio			3.58

		ppm	meq/L	lbs/A in
Calcium	Ca	238.6	11.93	54.23
Magnesium	Mg	47.2	3.94	10.73
Potassium	K	12.8	0.33	2.92
Sodium	Na	231.8	10.08	52.68
Iron	Fe	0.2		0.04

	ppm	meq/L	lbs/A in
Total Alkalinity	152.0		34.55
Carbonate	0.0	0.00	0.00
Bicarbonate	185.0	3.03	42.05
Chloride	753.0	21.36	171.14
Sulfate	83.1	1.74	18.89
Salt Concentration	1,728.6		392.87
Boron	0.05		
Cation/Anion Ratio		1.01	

Water 4

It is quite natural to look at the effect of irrigation water on soils and soil fertility, but let's step back and look at another aspect on plants that are under irrigation, especially those in greenhouses or some enclosed structure. The increase in humidity will impact nutrient uptake, especially for nutrients that are primarily xylem mobile, namely calcium and boron. These nutrients, and others to some degree, are moved up into the plant as a result of evapotranspiration. When the humidity rises, evapotranspiration is greatly reduced, along with the movement of calcium and boron. Calcium and boron are critical for cell wall integrity, and any interruption in the flow of calcium and boron can lead to weak stems, blossom end rot in fruit, and reduced growth.

The reduction in cell wall integrity lends itself to disease pressure such as powdery mildew surface fungal infections. Sucking insects may increase in population due to the increased ease in which they can penetrate and feed on the plant. Calcium is important in nitrogen metabolism, and a reduction in the level could lead to nitrate accumulation, enhancing the aforementioned problems. Boron is essential in sugar translocation, and with an increase in sugars in the plants, insects will be attracted, opening up the possibility for disease pressure.

Increasing air movement and controlling temperature is certainly beneficial for minimizing this problem, but in hot, humid climates and very cold times of the year, it will be challenging. Foliar feeding calcium and boron is an option, but it too will add to the humidity issue. Maximizing your levels of calcium and boron in you paste test will certainly be beneficial, especially while your plants are growing vegetatively and in the early stages of reproductive growth. Blocking uptake by excess sodium in the water or over-fertilizing potassium will only exacerbate the problem.

These problems can exist in field production during excessive wet periods and hot and humid conditions, but foliar feeding prior to or following these conditions can help reduce the stress place on the plants. Maintaining good soluble calcium levels in the soil through gypsum applications will be beneficial. Liming will also be helpful, provided the pH levels stay below 6.5. Don't forget — rainwater pH levels are around 6.6, so keeping your soil pH in the 6.2 range might be even more ideal.

Chapter 18

Future of Agriculture

After consulting in the field of agriculture for over forty years, I can say that I have seen tremendous progress in the protection of our most valuable resource — soils.

We have gone from moldboard plowing all of our production fields and suffering tremendous soil loss to farming virtually all of our production fields with conservation tillage and no-till farming practices. We are not at zero soil loss and probably never will be, but it is truly remarkable how far we have come. People just getting involved in agriculture sometimes become impatient with the speed of progress, but I feel that we are at least 80-85 percent there. The last 10-15 percent is always the most difficult and most expensive.

What is it going to take to continually improve soil quality and production? Money is the number one factor. When crop prices are good, it is amazing the willingness of farmers to experiment with new ideas and techniques. Poor crop prices continually keep farmers in survival mode and unwilling to take risks. In general, I see farmers as fearing the risk of loss more than risk of gain. I have been to organic conferences that tout the high price of organic products as a reason for the commercial farmers to move to organic practices. I seriously doubt if those price advantages could be maintained if a significant number of farmers shifted to organic production. It is simply supply and demand. It would be beneficial if both the commercial growers and the organic growers would move closer to the center. Until both organic and commercial growers are being paid for quality instead of volume, significant changes will be minimal at best.

If everyone switched to organic production whether I seriously question we could feed the population that exists now, let alone in the future. This comment is not meant to besmirch organic farmers in any way. I know this is a bold statement, and there are several reasons.

1. We just don't have nearly enough farmers to make this switch. Let's face it — organic farming is labor intensive.
2. Tillage will have to increase to achieve weed control, and getting commercial growers to go back to cultivating row crops is probably not going to happen, at least without a large incentive package. This increase in tillage will also increase the potential for soil loss.
3. Currently I seriously doubt if there is enough non-GMO seed in the pipeline to satisfy the demand.
4. This is probably true for fertilizer as well.

I don't really see a major shift in the number of commercial growers going into organic production — not only for the reasons mentioned above, but a three-year transition period would be financially crippling. There is a small shift in farmers planting into cover crops with reduced fertilizer and chemical inputs, but chemical control remains the backup plan. Unfortunately, due to current economic conditions, I do see the smaller and older farmer getting out of farming. This results in fewer farmers operating larger operations. Big does not necessarily mean bad, but it is the large operations that struggle with things like cover crops, doing away with fall nitrogen applications, and fertilizing just prior to planting. These large operations love no-till, which allows them to farm more ground. If no-till is resulting in stratification of the phosphorus and tillage is required to fix the problem, the larger operations will have more problems accomplishing this.

We need more diversity in our crops. Corn and beans only increase nematode, insect, and weed issues.

The onset of hemp will eventually bite into some of the corn and bean acres, but government involvement will drastically slow that process down.

I think we will see more urban farming, but that will only be for leafy greens and vegetables.

Technology is rapidly changing the way spraying, planting, and harvesting is being done. Farmers in general are on information overload. We have more data now on how the crop was planted — at what depth and rate as well as harvest moisture and yield, virtually by the square foot — but we know practically nothing about soil biology. Much of our soil chemistry research is from the 1940s and '50s. Most of this information is still valid today but not being fine-tuned for the changes in varieties and farming practices in today's agriculture.

Few people are using paste analysis, tissue analysis, and stalk nitrogen testing. There is no doubt that farmers of the future will need to be more technologically savvy from the equipment perspective, but they need to quit trying to micro-manage a soil system that is too complex and variable. Advancements in equipment technology is a wonderful thing, but it is not going to increase crop yields substantially. Yields are going to be dramatically increased through bio-engineering and balancing the soils. This will only be done when we use all the tools at our disposal, such as standard and paste tests, tissue analysis, stalk nitrogen, and available nitrogen testing — and do it on a zone basis. There still many farmers not even running the basic soil test, and those that do delegate it to the very people who are selling them fertilizer. Universities need to be turning out more independent consultants who practice the principles put forth in this book; however, that would require a huge change in attitude. The environmental challenges could be corrected with our current knowledge of soils, but this knowledge needs to be used and built upon if future generations are going to survive.

It is time to quite looking for the magic bullet and slapping Band-Aids on problems. It is time to get back to the basics of balancing the soil chemistry and improving soil organic matter and structure, and to leave the rest up to the Good Lord.

Notes

Acronyms

Acre furrow slice (AFS)
Anion (negatively charged ion)
Basic cation saturation ratio (BCSR)
Cation exchange capacity (CEC)
Cation (positively charged ion)
Evapotranspiration (ET)
Inductively coupled plasma (ICP)
Milliequivalents (mEq)
Organic matter (OM)
Strategic level of available nutrient (SLAN)
Total exchange capacity (TEC)
Two inches below and two inches off to the side (2 x 2)

Fertilizer Key

Ammonium nitrate (NH_4NO_3) 33-0-0
Ammonium sulfate ($(NH_4)_2SO_4$) 21-0-0
Anhydrous ammonia (NH_3) 82-0-0
Calcium carbonate ($CaCO_3$)
Calcium nitrate ($Ca(NO_3)_2 \cdot 4H_2O$)
Copper sulfate ($CuSO_4 \cdot H_2O$)
Diammonium phosphate (DAP) $(NH_4)_2HPO_4$ 18-46-0
Epsom salts, magnesium sulfate ($MgSO_4$)
Gypsum, calcium sulfate ($CaSO_4$)
K-Mag, sul-po-mag (potassium, magnesium and sulfur), 20-22% potassium, 11% magnesium and 20-22% sulfur
Monoammonium phosphate (MAP) $NH_4H_2PO_4$ 11-52-0
Muriate of potash, potassium chloride (KCl) 0-0-60
Potassium sulfate (K_2SO_4) 0-0-50
Urea ($CO(NH_2)_2$) 46-0-0

Index

A
Acre-furrow slices
  nutrient level calculations, 1
  organic matter content, 19
Adjusted Sodium Adsorption Rate, 205–206
ADP (adenosine diphosphate), 115, 180
ADT (adenosine triphosphate), 115, 180
Agriculture, future of, 217–219
Albrecht philosophy, of soil nutrient balance, 39. *See also* Basic Cation Saturation Ratio (BCSR)
Alfalfa
  boron tolerance, 101
  copper requirements, 109
  manganese tolerance, 105
  molybdenum application, 115
  nitrogen requirements, 109
  sodium application, 92
  sulfur affinity, 36
  zinc deficiency tolerance, 112
Alfalfa meal, 31
Algae, 45, 158
Aluminum, 5, 103, 128
Aluminum-phosphorus complex, 42
Amino acids, 23, 35, 114–115
Ammonia, 24, 29, 70
Ammonium, 24, 28, 43, 82
Ammonium acetate extraction test, 5, 6, 66, 166
Ammonium molybdate, 115
Ammonium nitrate, 31, 66
Ammonium sulfate, 31, 36, 63, 65, 141
Ammonium thiosulfate, 36
Anaerobic conditions, 37, 43, 173, 181, 201, 206, 208–209, 250
Anhydrous ammonium, 31
Anions, paste analysis, 128
Anthocyans, 43
Aphids, 53, 83
Apple orchards, heavy metal contamination, 118
Apple trees
  boron deficiency, 101
  calcium requirements, 52–53
  iron deficiency, 103
  manganese tolerance, 105
  nutrient removal rate, 26
Apricot trees, iron deficiency, 103
Aqua-ammonia, 31
Arsenic, 119
Asparagus, 92, 112
Auxins, 111
Azomite, 104, 107

B
Bacterial diseases, 83
Barley, 92, 97, 101, 110, 112
Base saturation
  of cations, 5
  variation, 4
Basic Cation Saturation Ratio (BCSR), 133
  calcium balance, 57–63
  contraindication for LEC soils, 86–87
  magnesium deficiency, 72, 74
BCSR. *See* Basic Cation Saturation Ratio
Beans
  calcium deficiency, 53
  chloride sensitivity, 97
  copper deficiency, 33
  copper requirements, 109
  diseases and pests, 218
  manganese tolerance, 105
  molybdenum deficiency, 33
  molybdenum fertilization, 115
  nitrogen application timing, 33
  nitrogen requirements, 109
  nutrient removal rate, 78
  phosphorus deficiency, 43
  sodium uptake inhibition, 92
  zinc deficiency tolerance, 112
Beets
  chloride sensitivity, 97
  molybdenum fertilization, 115
  nutrient removal rate, 26
  phosphorus deficiency, 43
  sodium uptake, 92
  zinc deficiency tolerance, 112
Berry bushes, chloride sensitivity, 97
Bicarbonates
  Adjusted Sodium Adsorption Rate, 205–206
  excessive levels, in HEC soils, 163, 165

iron deficiency and, 103
irrigation water content, 107, 165, 191, 206, 208, 210, 211, 212, 213
as nutrient deficiency cause, 206
paste analysis, 5, 123, 128
uptake in plant, 123
zinc mobility effects, 111
Bio-engineering, 219
Biosolids, 19, 20, 119
Blood meal, 31, 48, 104
Blossom drop, 111, 174
Blossom end rot, 52–53, 173, 214
Bone meal, 45, 46, 47–48, 54, 169
Borax, 101, 191
Borers, 117
Boric acid, 101
Boron, 5, 99–102
   irrigation water content, 206, 207, 210, 211, 212, 213
   for LEC soils, 137
   mobility in plants, 53, 99, 173
   paste analysis, 128, 144, 147
   role in plants, 99, 100, 214
   sources, 100–101
Boron deficiency, 99, 100, 101–102, 141
   in HEC soils, 150, 151, 152, 153, 154, 155, 156, 158, 159
   symptoms, 99, 100
Boron toxicity, 99, 101, 206
Broccoli, excessive nitrogen effects, 33
Brussel sprouts, chloride sensitivity, 97

C

Cabbage
   boron tolerance, 101
   nitrogen excess effects, 30, 33, 161
   nutrient removal rate, 26, 78
   sulfur affinity, 36
Cadmium, 119
Calcareous soils, 5, 76, 165–171, 173
   cobalt uptake, 116
   excessive nutrient levels, 173
   paste analysis, 66, 166, 168, 169
   soil tests, 166–167, 170
Calcium, 51–66
   base saturation, 60
   boron precipitation and, 100
   drought resistance and, 180
   effect on exchange capacity, 15
   foliar feeding, 52, 53, 176
   glyphosate chelation, 175
   in HEC soils, 57–66
   high soil content, 7
   ideal soil level, 51, 52
   importance, 51–53
   irrigation water content, 191, 208, 209, 210, 211, 212, 213
   low soil pH relationship, 16
   magnesium uptake effect, 70
   mobility in plants, 51, 52, 53, 173
   mobility in soil, 56
   paste analysis, 128
   potassium interactions, 87
   pounds per acre, 54, 56, 59, 74
   precipitation, 206, 208, 209
   role in plants, 51, 52–53, 214
   SLAN balance approach, 133–134
   in soil solutions, 83
   sources, 54–55, 56, 214
   standard soil test, 5, 53–54
   sulfated, 148
   tissue content, 53
   uptake from soil, 100
   water availability and, 180
Calcium borate, 100
Calcium carbonate, 54, 165–166, 168–169
   See also Calcareous soils
Calcium deficiency, 52–53
   in HEC soils, 57–66, 155–158
      in garden soil, 161–164
      with high magnesium and potassium levels, 161–165
      with high magnesium levels, 60, 62–63, 65–66, 155–158
      with high soil pH, 63–66
      with low soil pH, 59–63, 60–63
      with magnesium deficiency and high phosphorus, 158, 159
   in LEC soils
      with magnesium and potassium deficiencies, 73, 135–136, 138–139
      with magnesium deficiency, 54–56, 71, 73, 134–135
      paste test, 139, 144, 146, 147
      pounds per acre per milli-equivalent for, 74
      SLAN approach, 135, 138–139
   symptoms, 4, 51–53
Calcium lime, 54, 56, 71–72, 135
   for HEC soils, 60, 62–63
   for high pH soils, 63, 65–66
   for low pH soils, 59–63, 134
   magnesium uptake effects, 70–71
   pounds per acre, 63
Calcium-magnesium balance, 4, 39, 60
Calcium-magnesium silicate, 118
Calcium nitrate, 31, 54
Calcium phosphates, 42
Calcium-phosphorus complex, 42
Calcium silicate, 118
Calcium sulfate. *See* Gypsum
Calcium-to-manganese ratio, 107
Calcium toxicity, 51, 52

Cannabis, 82
Carbohydrates, 36, 37, 52, 92, 107, 114–115, 140
Carbonates, 103, 127, 205–206
   irrigation water content, 210, 211, 212, 213
Carbon-to-nitrogen ratio, 20, 27, 28
Carcinogens, 119
Carrots
   boron tolerance, 101
   manganese deficiency and toxicity, 106
   nutrient removal rate, 26
   sodium application, 92
   sodium uptake, 92
   zinc deficiency tolerance, 112
Cation(s)
   base saturation, 5
   milli-equivalents, 14
   parts per million/parts per acre conversion, 14
   paste analysis, 128, 129
   pounds-per-acre calculation, 11
   rebalancing approach, 133–134
   silicon products content, 118
Cation balance
   irrigation water effects, 207–213
   soil pH relationship, 15–16
Cation exchange capacity (CEC), 12
Cation-holding capacity. *See* Total exchange capacity
Cauliflower
   boron tolerance, 101
   chloride sensitivity, 97
   crop removal rate, 78
   excessive nitrogen effects, 33
   manganese tolerance, 105
   molybdenum fertilization, 115
Celery, 78, 115
Cereal crops, 53, 70, 82, 105, 107
Cereal rye, 25
Chelated iron, 104
Chelated manganese, 108
Chelated zinc, 113
Cherry tress, iron deficiency, 103
Chiseling, 60
Chloride
   application timing, 4
   crop sensitivity to, 97
   irrigation water content, 206–207, 210, 211, 212, 213
   molybdenum uptake effects, 114
   paste analysis, 5, 123, 128
   uptake in plant, 123
Chloride burn, 87
Chloride toxicity, 206–207
Chlorine, 95–97
   deficiency, 95, 96
   toxicity, 95
Chlorophyll, 35, 36, 43, 69, 70, 102, 104, 107, 110–111
Citrus/citrus trees, 112, 115
Clay/clay soils
   BSCR balance approach, 149
   buffer pH and, 15
   deflocculation, 208–209
   drought tolerance, 181
   exchange capacity, 13–14
   as HEC soils, 129
   irrigation effects on, 208–209
   permeability problems, 206
   potassium fixation, 83, 85, 86
   soil pHc, 206, 209, 210, 211, 212
   structure and granulation, 179
   water availability, 180
   zinc mobility in, 111
Clean Air Act, 17, 37
Clovers, 43, 92, 105
Coastal soils, sodium exchange capacity content, 13
Cobalt, 115, 116
   deficiency, 115
   toxicity, 115
Cobalt nitrate, 116
Cobalt sulfate, 116
Cold tolerance, 107, 109, 111, 117, 140
Compost
   application time, 136–137
   conversion to humus, 19
   as Modified Growing Media component, 187–188, 189, 190
   as nitrogen source, 31
   paste analysis, 188, 190
   as phosphorus source, 46
Copper, 108–110
   allowable cumulative loading rates, 119
   deficiency, 33, 108–109
   mobility in plants, 108, 109
   molybdenum uptake effects, 114
   nickel uptake effects, 120
   paste analysis, 128, 144, 147
   soil test, 5
   sources, 110
   toxicity, 108, 109–110
   uptake in plant, 123
Copper chelates, 110
Copper oxide, 110
Copper ppm/nitrogen percent, 109
Copper sulfate, 37, 109, 110, 137, 140, 151, 158, 176
Coral-based soils, 5, 7, 14–15, 76, 165
Corn
   boron tolerance, 101
   chloride sensitivity, 97

diseases and pests, 218
foliar feeding, 175
fungicides, 37
nitrogen deficiency, 30
nitrogen fertilizer application timing, 32
nitrogen monitoring, 28–29
nitrogen requirements, 25, 26, 32–33
nitrogen-to-potassium ratio, 82
nutrient removal rates, 26, 78, 84, 86, 141
physiological maturity (black layer), 29–30
potassium requirements, 87
side-dress nitrogen utilization, 28–29
sodium uptake inhibition in, 92
sulfur affinity, 36
sulfur deficiency, 36
zinc deficiency tolerance, 112
Corn gluten meal, 31
Corn stalk analysis, 29–30
Corporate agriculture, 218
Cotton, 92, 105
Cover crops, 218
for erosion control, 44, 45
nitrogen loss effects, 25
as nitrogen source, 27
phosphorus assimilation effects, 137
phosphorus stratification and, 43–44
as potassium source, 87
urea-based protein content, 28
water availability effects, 181
Crab meal, 31, 45
Crop diversity, 218
Crop removal charts, 47
Crop removal rates, 26
See also nutrient removal rate *under specific crops*
in LEC soils, 135
Crop residue, 27, 87, 181
Crossover (calcium-magnesium silicate), 118
Cucumbers, 26, 92, 97, 105, 117

D
DAP fertilizer, 137, 142
Denitrification, 57, 173, 181
Di-ammonium phosphate, 31, 46
Dicots, 100, 102, 107, 114, 175
Die back, 111
Diffusion, of nutrients, 179
Direct root intercept, of nutrients, 102, 109, 123, 158, 179
Disease resistance, 140
Disease susceptibility, 37, 83, 214
Dolomite lime, 54, 56, 76
excessive, 150–151

for low pH soils, 59–60, 133–135
as magnesium source, 56, 71, 74, 76, 134–135
for Modified Growing Media, 191
Double-spreading, of fertilizer, 2
Drainage, 133
effect on soil sulfur levels, 38
of excessive magnesium, 63
leaching and, 161
of Modified Growing Media, 200–202
of raised beds, 200–202
sulfur levels effects, 129
Drought/drought stress, 44, 70, 117, 140, 165, 180–181
Dry soils
calcium availability, 56
molybdenum deficiency and, 114
nutrient availability, 180
nutrient concentration, 165
paste analysis results, 129
phosphorus availability in, 42–43, 44, 45
potassium uptake, 83
soil microorganisms, 44
zinc uptake, 111

E
Earthworm channels, 44–45
EDTA chelates, 103–104
Electrical conductivity (EC), 205
Emitters, 208
Environmental laboratory analyses, 118–119
Enzyme system catalysts, 103, 104, 107, 109
EPA 503 metals analysis, 119
Epsom salts. See Magnesium sulfate
Estimated nitrogen release (ENR), 26, 27
Evaporation rate, 181
Evapotranspiration, 53, 173–174, 214
Exchange capacity, 133
inaccuracies, 14–15, 76
measurement, 14
true, 14
Extraction solutions
See also Mehlich extraction solution
differences among, 6–7

F
Feather meal, 31, 48
Feed analysis, 150
Ferrous oxide, 104
Ferrous sulfate, 104
Fertilizers
application timing, 3–4
nutrient concentrations, 31
organic, 218

Field capacity, 179, 180
Fish emulsions, 176
Fish meal, 31, 46
503 metals analysis, 119
Foliar feeding, 173–176, 182
   penetrants, 175
   reasons for, 173–174, 176
   trace elements, 148–149
   weather conditions affecting, 175, 214
Free carbonates, 6, 7, 14–15, 66, 76–78, 100, 150–151
Frost damage, 83
Fruit trees
   See also Apple trees; Peach trees; Pear trees
   boron deficiency, 101–102
   chloride sensitivity, 97
   iron deficiency, 103
   zinc deficiency tolerance, 112
Fulvic acid, 169
Fungal diseases, 83, 174, 214
Fungicides, 37, 110
Fusarium, 174

G

Gardens
   calcium, magnesium, and potassium deficiencies, 55–56
   excessive nutrient levels, 161–165
Germination
   ammonia toxicity to, 24
   boron requirements, 99
   calcium requirements, 51, 52
   chloride effects on, 97
   chloride toxicity to, 97, 157
   in copper deficiency, 109
   in growth media mix, 194
   magnesium requirements, 69
Glyphosate, 65, 106, 111, 120, 174
Granubor calcium borate US borax, 101
Grapes, 112
Grasses
   boron requirements, 100
   magnesium requirements, 70
   manganese deficiency, 107
   molybdenum uptake, 114
   potassium uptake, 88
   silica accumulation, 117
   zinc deficiency tolerance, 112
Grass tetany, 88
Greenhouses, 214, 217
Growing mediums
   See also Modified Growing Media (MGM)
   silicon fertilization, 117
Gypsum, 36
   calcium percentage of, 148, 150
   for HEC soils, 63, 65, 66, 76, 150, 157
   for high organic matter soil, 161, 164
   for high pH soils, 65, 133–134
   as irrigation water additive, 209
   for LEC soils, 54, 148
   magnesium balance and, 38, 39, 71, 76, 88, 148, 150, 157, 164
   for Modified Growing Media, 191
   pounds per acre, 63
   in sulfur deficiency, 161

H

Hail damage, 174
Harsco company, 118
Heavy metals, 118–120
HEC soils
   calcium and magnesium imbalance, 150–153
   calcium deficiency with magnesium with potassium excesses, 161–162, 164–165
   calcium-magnesium imbalance, 150–153
   magnesium deficiency, 61
   potassium deficiency, 61
HEC soils. *See* High-exchange capacity soils
Hemp, 218
Heptonates, 103–104
Herbicides, 174
High-exchange capacity (HEC) soil balance, 124
High-exchange capacity (HEC) soils
   See also Basic Cation Saturation Ratio (BCSR)
   anaerobic conditions, 57
   base saturation numbers, 133
   calcareous soils, 165–171, 173
   calcium and magnesium balance, 59–60, 61–66, 72, 74–78, 150–153
      calcium base saturation, 63, 65
      calcium deficiency, 59–66
      high pH soils, 63–66
      lime applications, 59–60, 63, 65–66
      low pH soils, 59–63
      magnesium base saturation, 60
      magnesium deficiency, 59, 72, 74–78
      magnesium excess, 60, 62, 63, 65–66
   cation content, 129
   clay content, 57, 58, 129
   lime applications, 16, 59–60, 63, 65–66
   nitrogen loss, 57
   organic matter content, 58, 129
   paste analysis, 128, 129, 154, 156, 160, 163, 168

potassium balance, 81, 87–88
silt content, 57, 58
trace elements, 151, 152, 153, 154, 155, 156
Hoop houses, 161
Hops, 112
Humates, 169
Humic acid, 169, 171
Humid conditions, 53, 99, 173–174, 175, 214
Humus, 13, 14, 17, 20
Hydrogen, 14, 15, 59, 74
Hydrogen sulfide, 36

I
ICP (Inductively Coupled Plasma) unit, 6, 11, 13, 15, 76
Illite, 13
Insect damage, 214
Iron, 102–104
  foliar feeding, 103
  irrigation water content, 208, 210, 211, 212, 213
  mobility in plants, 102, 103
  molybdenum uptake effects, 114
  paste analysis, 128
  soil test for, 5
  sources, 103–104
  tissue sampling, 103
  uptake in plant, 123
Iron chelates, 103–104
Iron chlorosis, 103I
Iron deficiency, 102–104, 107
Iron lignosulfates, 104
Iron-phosphorus complex, 42
Iron polyflavonoids, 104
Iron sulfate, 114, 171
Iron toxicity, 102
Irrigated soils, sodium exchange capacity, 13
Irrigation
  drip, 165, 207, 208
  in enclosed structures, 214
  overhead, 207
Irrigation water
  acidification, 208
  Adjusted Sodium Adsorption Rate, 205–206
  adverse effects on soil, 206
  bicarbonates content, 107, 185
  boron content, 206, 207
  cation balance effects, 207–213
  chloride content, 96, 206–207, 210, 211, 212, 213
  leaching effect, 207, 208, 211
  for Modified Growing Media, 191
  paste analysis, 123, 127–128

Sodium Adsorption Rate (SAR), 205, 207, 209

K
Kaolinite clay, 13, 111
Kelp, 107, 176
Kernel abortion, 174
KMag, 36
  application prior to irrigation, 208
  for calcareous soils, 76, 78, 169
  for high soil pH correction, 133–134
  for magnesium balance, 56, 71–72, 134, 141, 191
  for Modified Growing Media, 191
  for potassium balance, 71–72, 88
  potassium-to-magnesium ratio, 82
  pounds per acre, 72
  as sulfur source, 72
  zinc soil availability and, 112

L
Lawns
  grass clippings as nutrient source, 84, 140
  potassium deficiency, 140–144
Leaching, 161, 173
  of bicarbonates, 165
  irrigation water-related, 207, 208, 209, 211
  magnesium sulfate (Epsom salts), 39
  of molybdenum, 182
  of nitrates, 38
  of nitrogen, 181
  in saturated soils, 181
  of sulfates, 38
Lead, 119–120
Leaf burn, 176, 193, 194
Leaf-hoppers, 83, 117
Leaf scorch, 109
LEC soils. *See* Low-exchange capacity soils
Legumes
  calcium deficiency, 53
  growing period, 28
  magnesium requirements, 70
  nitrogen deficiency, 35–36
  as nitrogen source, 25, 28, 33
  nodule bacteria, 35
  nodule formation, 182
  sulfur deficiency, 35–36
Lettuce
  boron tolerance, 101
  chloride sensitivity, 97
  copper sensitivity, 110
  manganese tolerance, 105
  molybdenum fertilization, 115
  nitrogen requirements, 161

nutrient removal rates, 78
sodium uptake inhibition in, 92
zinc deficiency tolerance, 112
Lignins, 164
Lignosulphonates, 103–104
Lignum, 111
Lime, 16, 54, 56, 214
  See also Calcium lime; Dolomite lime
  application timing, 3
  burnt, 54
  dissolution in extraction solution, 76
  effect on exchange capacity values, 7, 14–15
  excessive, 100, 107
  high-calcium. *See* Calcium lime
  low soil pH relationship, 16
  as magnesium source, 54, 56
  pH, 14–15
  precipitation, 206
  rainwater pH and, 17
  soil stratification, 112
  solubility, 76, 150–151, 206
  zinc availability effects, 112
Liquid Polyphosphate (10-34-0), 46
"Little leaf," 111
Little leaf syndrome, 120
Loamy soils, sodium application, 92
Lodging, 43, 53
Logan Labs, Ohio, 1, 11, 37
  Modified Growing Media standard soil analysis, 185–188
Low-exchange capacity (LEC) soils, 134–149
  See Sufficiency Level of Available Nutrients (SLAN) approach
  calcium and magnesium balance, 54–57
    calcium-to-magnesium ratio, 71
    cation content, 129
    lime applications, 56, 134–135
    magnesium deficiency, 54–56, 71–72, 73, 134–135
    magnesium uptake, 70–71
  calcium deficiency, 55, 138
  fertilizer application timing, 3, 4
  field location, 138
  lawns, 140–144
  lime applications, 16, 56
  magnesium deficiency, 138
  organic matter content, 134
  paste analysis, 128, 129, 139, 140, 146, 147
  phosphorus deficiency, 136–137, 141–142
  potassium deficiency, 71, 72, 73, 86–87, 135–136, 138–139, 140–141, 143–144
  samples to balance, 134–149
  soil balancing examples, 134–149
  standard soil reports for, 134–149
  sulfur deficiency, 61, 63, 140–141, 143, 144, 147

M

Magnesium, 69–78
  crop removal rates, 78, 141
  effect on exchange capacity values, 15
  as enzyme catalyst, 107
  foliar feeding, 176
  glyphosate chelation, 175
  in HEC soils, 150–151
  interaction with potassium, 82
  irrigation water content, 191, 208, 210, 211, 212, 213
  low soil pH relationship, 16
  mobility in plants, 69
  for Modified Growing Media, 191
  paste analysis, 128
  phosphorus uptake effects, 142
  potassium interactions, 87
  pounds-per-acre calculations, 59, 74, 76, 135
  precipitation, 206, 208
  soil balance, 133–134
  in soil solutions, 83
  soil test for, 5
  sources, 56, 71, 134–135, 141
  sulfated, 148
  time required to balance, 151
  tissue sampling, 31
  underestimation, 7
  uptake from soil, 100
Magnesium carbonate, 63, 150–151, 169
Magnesium deficiency, 69
  with calcium and potassium deficiencies, 155–158
  in HEC soils, 59, 72, 74–78
  in LEC soils, 54–56, 71–72, 73, 134–135, 138–139
  with low soil pH, 60, 62–63
  paste test findings, 139, 141, 144
  SLAN balancing approach, 134–135, 138–139
  soil reports, 72, 74–76
  sulfates-related, 38
Magnesium excess, 57, 82
  in calcareous soils, 168, 169
  with calcium deficiency, 65–66, 155–158
  with calcium deficiency and potassium excess, 161–165
  drainage, 63
  in HEC soils, 60, 62–63, 65–66, 150–151, 153–154

Magnesium oxide, 71
Magnesium oxysulfate, 71
Magnesium silicate, 118
Magnesium sulfate (Epsom salts), 36, 38, 39, 56, 71, 133–134, 148, 164, 191
Magnesium toxicity, 69
Manganese, 104–108
  as enzyme catalyst, 104, 107
  irrigation water effects on, 208
  magnesium requirements and, 70
  mobility in plants, 104, 106–107
  molybdenum uptake effects, 114
  paste analysis, 128
  soil test for, 5
  sources, 107–108
  tolerance, 105
  toxicity, 104–105, 106
Manganese carbonate, 108
Manganese deficiency, 104, 105–106, 141
Manganese oxide, 108
Manganese sulfate, 108, 137, 176
Manure, 32
  application timing, 3, 4
  beef, 46
  as calcium source, 52, 54
  as copper source, 110
  fall application, 3
  hog, 46, 110, 112
  as nitrogen source, 136
  organic matter content, 112
  phosphates content, 112
  phosphorus or potassium imbalance and, 32
  as phosphorus source, 136, 142
  as potassium source, 86, 136
  poultry, 46, 54, 136
  urea-based protein content, 28
  zinc content, 112
Manure analysis, 32
MAP fertilizer, 137, 142, 157–158
Marl-based soils, 14–15
Mass flow/diffusion, of nutrients, 123, 179
Mehlich extraction solution, 5, 66, 166
  advantages, 6–7
  pH, 5, 7, 76, 166
Mehlich 3 extraction solution, 4, 6, 119, 187
  pH, 66, 67
Melons, 97, 117
Mercury, 119
Metal-chelating agents, 174
Milli-equivalents, 15
Mineral Nutrition and Plant Diseases (Datnoff, Elmer, and Huber), 120
Mineral soils, metals fixation in, 119

Minimum tillage, 43–44, 63, 65
Modified Growing Media (MGM), 185–202
  base mix analysis, 185–188
  bulk density number, 185–186, 187
  drainage, 200–202
  flushing of, 194, 198, 199–200
  formulation, 186–187
  guidelines, 192
  magnesium-to-potassium ratio, 187
  paste analysis, 185, 192, 194, 197, 199
  pH, 187, 191
  standard soil reports, 189, 193–194, 195–196, 198
Moisture stress, foliar feeding and, 175
Moldboard plowing, 45, 60, 217
Molybdenum, 113–115, 119
  manganese mobility and, 107
  nutrient removal rate, 114
  sources, 115
  tissue sampling, 31
Molybdenum deficiency, 33, 113, 114
Molybdenum toxicity, 113
Mono-ammonium phosphate, 31, 46, 169
Monocots, 100, 175
Montmorillonite, 13, 111
Morgan extraction solution, 6
Muck fields, 105–106
Mulch, 164
  plastic, 207
Muriate of potash, 135
Muskmelons, nutrient removal rates, 26

N

Nickel, 31, 119, 120
Nickel deficiency, 120
Nitrate nitrogen, 24, 28, 29, 70, 114
Nitrate reductase system, 113
Nitrates, 35, 37, 38, 100, 109
Nitrogen, 23–33
  acid-forming, 107
  application, 25, 32–33
  atmospheric, 35
  background levels, 28, 38
  banded, 28–29
  breakdown in soil, 24
  calcium demand and, 53
  calcium uptake effects, 52
  corn stalk analysis, 29–30
  crop removal rates, 25–27
  crop requirements, 25–27, 158
  denitrification, 57, 173, 181
  excessive, 30
  ideal soil levels, 23
  iron deficiency and, 103
  loss, 29

mobility in plants, 23, 30–31
overfertilization with, 117
release from organic matter, 20
side-dressing, 28–29
soil nitrogen test, 28–29
sources, 31, 97, 136, 141, 158, 161
sulfur fertilizers use with, 36
as surface ground water contaminants, 29, 30
time of application, 29
types, 31–32
variation in requirements for, 23–24
in wet soils, 181–182
Nitrogen deficiency
chloride-related, 96
symptoms, 23
tissue sampling, 30–31
Nitrogen degradation cycle, 24
Nitrogen-fixing bacteria, 114
Nitrogen reductase enzyme system, 35, 109
Nitrogen test, 5
Nitrogen-to-potassium ratio, 82
Nitrogen-to-sulfur ratio, 36
Nitrogen toxicity, 23, 161, 214
Nitrous oxide, 24
non-GMO seeds, 218
No-till agriculture, 217
adverse effects, 218
as contraindication to rock phosphate, 48
fertilizer application timing, 3
limestone application in, 15
phosphorus stratification in, 43–44
zinc availability in, 112
NPK (nitrogen, phosphorus, potassium), 25, 26, 48, 86
Nutrient availability
effect of irrigation water on, 207
effect of soil pH on, 17
weather and, 179
Nutrient balance
See also specific nutrients
for specific crops, 4
Nutrient levels
historical comparisons, 1
postharvest, 3
Nutrient stratification, 1–2
Nutrient uptake
in humid conditions, 173–174, 214
through direct root intercept, 123–124, 179
through mass flow or diffusion, 123, 179

O

Oats, 92, 101, 105, 112

Ohio, No-Till Farmer of the Year recipient, 44
Olsen extraction solution, 166
Onions
boron tolerance, 101
chloride sensitivity, 97
foliar feeding, 175
nutrient removal rates, 26, 78
sodium uptake inhibition in, 92
sulfur affinity, 36
zinc deficiency tolerance, 112
Organic acids, 208
in paste analysis, 127
Organic agriculture
calcareous soil, 169
chloride deficiency, 96
limitations, 217–218
Modified Growing Media, 191
phosphorus sources, 46, 47–48, 136, 191
potassium sources, 85, 86, 135, 136
sea-based foliar feeding products, 176
Organic fertilizers, timing of application, 3
Organic matter, 5
See also Humus
benefits for soil, 179
BSCR approach and, 58
buffer pH and, 15
drought tolerance and, 181
excessive nitrogen, 47
exchange capacity, 13, 14
gypsum requirements and, 161–164
in HEC soils, 58, 129
high soil content, 83
iron deficiency and, 103
in LEC soils, 134
low soil content, 70
measurement method, 17–19
as nitrogen source, 20, 30, 158, 161
silica deficiency and, 117
soil micro-organisms as, 19, 20
soil test levels, 17–19
zinc soil mobility effects, 111
Organic produce, prices, 217
Organic products, for soil balance, 134
Ortho-phosphorus products, 141–142
Oxidative/reduction reactions, 107

P

Palms, chloride sensitivity, 97
Parakeratosis, 112
Parts per million (ppm), conversion to pounds per acre, 1, 11, 14, 149–150
Paste analysis, 5, 123–130, 139, 219
bicarbonates, 5, 123, 128, 163, 165
boron, 102, 144, 147, 151

buffering capacity, 129
calcareous soils, 66, 76, 78, 166, 168, 169
calcium, magnesium, and potassium balance, 139, 161
calcium and magnesium balance, 16, 76, 147
calcium deficiency, 53–54, 56, 144, 147
calcium deficiency with magnesium and potassium excesses, 161, 163, 164–165
calcium levels, 16
calcium with magnesium and potassium deficiencies, 160
chlorine, 97
compost, 188, 190
copper, 144, 147
definition, 123
foliar feeding and, 214
garden soil, 161, 163, 164
HEC soils, 124, 154, 156, 158, 160, 163, 168
   calcium and magnesium imbalance, 154
   guidelines, 128
   magnesium deficiency, 76
iron, 151
irrigation water, 123, 127–128
LEC soils, 133, 139, 146, 147
   calcium, magnesium, and potassium balance, 139
   calcium and magnesium balance, 76
   guidelines, 128
   in lawns, 140, 141, 144
   phosphorus deficiency, 141, 144
   potassium deficiency, 141, 144
   solubility levels, 133
   sulfur deficiency, 141, 144
   trace elements deficiency, 141, 144
manganese deficiency, 151
Modified Growing Media, 185, 192, 194, 197, 199
nitrogen, 160
organic acids, 127
parameters and guidelines, 5, 128
phosphorus, 128, 141, 144, 157
pH range, 16
plan for balancing, 134
potassium, 64, 831
sod farm, 124, 126–127
sodium, 128
soil pH, 16, 127–128
soil test findings comparisons, 67, 124–127, 134, 141, 147, 158
of soil test samples, 147
soluble salts, 5, 123, 128, 129
   for specialty crops, 5
   sulfur, 37, 128, 144, 146, 147
   tissue analysis and, 124, 147
   trace elements, 128, 141, 144, 146, 169
   zinc deficiency, 144, 147
Peach trees
   boron deficiency, 101–102
   iron deficiency, 103
   nutrient removal rates, 26
Pear trees, 101, 103
Peas
   boron tolerance, 101
   chloride sensitivity, 97
   nutrient removal rates, 26
   zinc deficiency tolerance, 112
Peppers, 26, 30, 173
Permeability, factors affecting, 205–206
pH, of soil. *See* Soil pH
Phosphates, 103, 114
Phosphorus, 41–48
   in ADP conversion, 180
   availability in soil, 141–142
   banding, 45
   complexes, 42
   crop removal rate, 26, 47, 136
   drought resistance and, 180
   elemental, conversion to P2O5, 149–150
   excess, with calcium and magnesium deficiencies, 158, 159
   excessive levels, 173
   ideal soil level, 41, 47
   irrigation water effects on, 208
   mobility in plants and soil, 41, 42–43, 44, 70, 100, 137
   as P_2O_5, 142, 149–150
   paste analysis, 128, 141, 144, 157
   pounds-per-acre-calculations, 11, 142, 149–150, 193–194
   precipitation, 16, 42–43, 208
   in saturated soils, 181
   SLAN balance approach, 134
   soil content, 32
   soil stratification, 42, 43–45, 218
   soil test for, 5
   soluble, banded, 181
   sources, 46, 47–48, 136, 142, 169
   toxicity, 41
   zinc interaction, 111
Phosphorus deficiency, 43, 100, 115, 136–140
   in LEC soils, 136
   paste test findings, 141, 144
   symptoms, 41
   tissue sampling for, 31
Photosynthesis, 70, 83, 100, 107
Physical testing, 5

Phythium, 174
Phytin, 42
Phytophthora, 174
Pin oak trees, 169, 171
Plant canopy, water availability effects, 181
Potassium, 81–88
  application methods and timing, 87, 157–158
  calcium uptake effects, 52
  crop removal rate, 26, 84, 86, 88, 136
  drought resistance and, 180
  foliar feeding, 176
  interaction with magnesium, 88
  irrigation water content, 208, 210, 211, 212, 213
  magnesium uptake effect, 70
  mobility in plants, 81
  paste analysis, 128
  phosphorus uptake effects, 142
  pounds-per-acre calculations, 59, 74, 85, 194
  precipitation, 208
  role in plants, 83
  in soil solutions, 83
  soil stratification, 88
  soil test for, 5
  sources, 85, 86, 97, 136, 140
  tissue sampling for, 31
  types, 31
  underestimation, 7
  uptake from soil, 82, 100, 181
  water availability and, 180
Potassium chloride, 86–87, 96, 97, 107, 136, 140, 141, 157
Potassium deficiency, 6, 83
  with calcium and magnesium deficiencies, 155–158
  in lawns, 140–141, 140–144, 143, 144
  in LEC soils, 71, 72, 73, 135–136, 138–139, 140–141, 143–144
  pounds per acre per milli-equivalent for, 74
  SLAN approach, 135–136
  soil reports, 55, 138
  symptoms, 4, 41, 43
  water utilization in, 175
Potassium excess
  with calcium deficiency and magnesium excess, 161–165
  with high soil pH, 63, 64
Potassium magnesium sulfate. Se KMag
Potassium nitrate, 31
Potassium silicate, 118
Potassium sulfate, 36, 38, 39, 86–87, 136, 140, 141, 157, 158, 176
Potassium sulfate magnesium, 38
Potassium-to-magnesium ratio, 82
Potassium toxicity, 41, 82
Potatoes
  boron tolerance, 101
  chloride sensitivity, 97
  copper sensitivity, 110
  magnesium requirements, 70
  manganese tolerance, 105
  nutrient removal rates, 26, 86–87
  phosphorus deficiency, 43, 136–137
  potassium requirements, 86–87
  starch content, 96
  zinc deficiency tolerance, 112
Pound-per-acre calculations, 1
Pounds-per-acre calculations, 149
  calcium, 54, 56, 59, 74
  calcium lime, 63
  converted from parts per million, 1, 11, 14, 149–150
  gypsum, 63
  hydrogen, 59, 74
  KMag, 72
  magnesium, 59, 74, 135
  MAP, 137
  nitrogen, 29
  P_2O_5, 142, 149–150
  phosphorus, 11, 193–194
  potassium, 59, 74, 85, 194
  sodium, 59, 74, 92, 194
  sulfur, 38–39
Powdery mildew, 116, 117, 214
Protein content, of plants, 37, 92, 114–115
Pumpkins, 117

R

Radishes, 36, 97
Rainfall
  boron availability and, 53
  calcium availability and, 53, 56
  as cation ions loss cause, 129
  excessive, 181–182
  foliar feeding and, 175
  low, 133
  phosphorus run-off in, 44–45
  soil balancing and, 133
Rainwater
  as chloride source, 96
  for irrigation, 209, 210
  pH, 17, 214
Raised beds, drainage, 200–202
Raspberries, 101, 102
Rhizoctonia, 174
Rice, 105, 117
Rock dusts, 3, 107, 148–149
Rock phosphate, 3, 42, 169
  application time, 136–137

as calcium source, 54
Florida, 48
as manganese source, 107
as phosphorus source, 46, 47–48, 150
Tennessee brown, 48
Roots
  calcium requirements, 52
  nutrient uptake, 102, 103, 109, 123–124
  soil sampling of, 19–20
Rot rots, 82
Rye, 112

S

Salt burn, 82
Salt-tolerant plants, 129, 209
Salt toxicity, 96
Sandy soils
  boron deficiency, 100
  cobalt uptake, 116
  DAP (18-46-0) products in, 137
  exchange capacity, 13–14
  MAP (11-52-0) products in, 137
  nutrient- and water-holding capacity, 179
  silica availability, 117
  sodium application, 92
  water availability, 179, 180
Sap analysis, 174
Saturated soils, as iron deficiency cause, 103
Selenium, 119, 120
Serpentine soils, 165–166
Shikimate enzyme system, 174
Side dress trenches, nitrogen monitoring, 28–29
Silicic acid, 116
Silicon, 107, 116–118
Silt, exchange capacity, 13–14
SLAN. *See* Sufficiency Level of Available Nutrient (SLAN) approach
Sod
  organic matter tests, 19–20
  paste analysis/soil test comparison, 124–127
Sodium, 91–92
  in ADP conversion, 180
  as exchange capacity component, 13
  irrigation water content, 208–209, 210, 211, 212, 213
  paste analysis, 128
  pounds per acre, 59, 74, 92, 194
  soil test for, 5
  sources, 92
Sodium Adsorption Rate (SAR), 205, 207, 209
Sodium-based extraction solutions, 6
Sodium chloride, 92
Sodium deficiency, 74, 91
Sodium molybdate (dry or liquid), 115
Sodium nitrate, 31, 92
Sodium toxicity, 91, 209
Soil balancing examples, 133–171
  HEC soils, 149–171
  LEC soils, 134–149
Soil chemistry, 219
Soil compaction, 3, 82, 88, 103, 180, 181
Soil depth, constant level, 1
Soil erosion, 44, 45
Soil granulation, 19–20
Soil laboratories
  differences in procedures among, 7
  selection of, 4, 7
  soil samples for, 2–3
Soil micro-organisms
  in calcareous soils, 169
  calcium-to-nitrogen ratio, 30
  in dry soil, 44
  as Modified Growing Media component, 188
  nitrogen breakdown by, 24, 27
  in organic matter tests, 19, 20
  phosphorus availability and, 42
  in wet soils, 181–182
Soil nitrogen test, 28–29
Soil pH, 5, 15–17, 133
  buffer pH, 15, 16–17
  cation balance relationship, 15–16
  of clay (pHc), 206, 209, 210, 211, 212
  definition, 15
  effect of nitrogen fertilizers on, 31
  heavy metal availability effects, 120
  high, 5
    balance, 133–134
    calcium-magnesium balance and, 63–66
    cation rebalancing approach, 133–134
    in HEC soils, 155–157
    in minimum tillage, 63, 65
  low
    balance, 134
    calcium-magnesium balance and, 60–63
    cation rebalancing approach, 134
    in LEC soils, 134
    magnesium deficiency, 60, 62–63
  manganese availability and, 107
  measurement, 15
  molybdenum deficiency and, 114
  on paste analysis, 127–128
  prior to irrigation, 208
  rainwater pH effects, 214
  silicon products effects on, 118

sulfur's effect on, 36
zinc soil mobility effects, 111
Soil pH charts, 16, 17
Soil probes, 2, 3
Soil sample depth
  importance, 1–2
  listed on soil report, 11, 12
  organic matter content and, 20
  of side-dress trenches, 29
  use in nutrient level calculations, 1, 11
Soil sampling
  goal, 4
  method, 2–3
  random collection if samples, 3
  recommended number of samples, 2–3
  recommended timing, 3–4
  sample analysis procedure, 5–8
Soil stratification, 42, 43–45, 88, 112
Soil structure
  irrigation effects on, 207–209
  water availability and, 180
Soil surfactants, 165, 181, 185
Soil texture triangle, 57–58
Solubility balance, paste analysis for. *See* Paste analysis
Soluble salts, 205, 208, 209
  in Modified Growing Media, 186–187, 194, 195–199, 200–201
  paste analysis, 5, 123, 128, 129
  uptake in plant, 123
Solubor US borax, 101
Soybeans
  boron deficiency, 100
  fungicides, 37
  glyphosate-related manganese deficiency, 106
  iron uptake, 102
  legume cover crops and, 28
  nutrient removal rates, 26, 114
  trace elements foliar feeding, 176
Spades, use for soil sampling, 2–3
Spinach, 92, 97, 112, 115
Stalk nitrogen testing, 29–30, 219
Standard soil tests and reports, 5
  See also High-exchange capacity (HEC) soils; Low-exchange capacity (LEC) soils; Paste analysis; *names of specific soil nutrients*
  calcareous soils, 166, 167, 170
  comparison with paste analysis, 124–127
  compost, 189
  evaluation guidelines, 133–134
  gardens, 161–165
  lab number, 11, 12
  lawns, 140–144
  Modified Growing Media (MGM), 185–188, 189, 193–194, 195–196, 198
  plan for balancing, 133–134
  sample depth, 11, 12
  sample ID, 11, 12
  sample location, 11, 12
  selection of tests, 5
  total exchange capacity, 12, 13–15
Starch, 83, 96
Straw, as mulch, 164–165
Strawberries
  boron tolerance, 101
  chloride sensitivity, 97
  manganese tolerance, 105
  sodium uptake inhibition in, 92
Sucking insects, 117, 214
Sufficiency Level of Available Nutrient (SLAN) approach, 57, 133, 134–149
  calcium deficiency, 54, 56, 57, 135, 138–139
  lawns, 140–144
  low soil pH, with calcium, magnesium, and potassium deficiencies, 134–140
  magnesium balance, 71–72, 74–76, 134–135, 138–139
  phosphorus balance, 134, 136–140
  potassium deficiency, 86–87, 135–136, 138–139, 140–143
  sulfur balance, 134, 140, 141, 143, 144
  trace elements balance, 134
Sugar beets, 70, 110
Sugars, plant content, 96, 100, 107, 114–115, 214
Sulfated minerals, 37
Sulfates, 36, 38
  application timing, 4, 38
  irrigation water content, 210, 211, 212, 213
  molybdenum uptake effects, 114
Sulfur, 35–39
  atmospheric, 37
  determination of amount needed, 38–39
  effect on excess soil nitrogen, 83–84
  elemental, 3, 36, 38, 39
  ideal soil levels, 35, 38–39
  mobility in plants, 35
  paste analysis, 37, 128, 129, 144, 146, 147
  pounds-per-acre calculations, 38–39
  SLAN balance approach, 134
  soil test for, 5
  sources, 36–37, 38, 140
  types, 36
Sulfur-affinity crops, 36, 38
Sulfur deficiency, 83, 118
  in HEC soils, 153, 154, 155, 156, 157, 159, 160

increased incidence, 37
in LEC soils, 61, 63, 140–141, 143, 144, 147, 148
paste test, 144, 147
SLAN approach, 140–141, 143, 144
standard soil test findings, 145, 147
symptoms, 35–36
Sulfuric acid, 208
Sulfur toxicity, symptoms, 35
sul-po-mag. *See* KMag

T
Take-all disease, 107
Technology, effect on agricultural practices, 218–219
TEC. *See* Total exchange capacity
Temperature/humidity index, 175
Thiobacillus, 39
Tillage practices
  See also Minimum tillage; No-till agriculture
  changes in, 217
  effect on soil profile, 1–2
Timothy grass, 88, 105
Tissue analysis, 219
  aluminum, 103
  calcium, 148
  copper, 109, 158
  iron levels, 103
  magnesium, 148
  Modified Growing Media, 189, 195–196, 198
  nitrogen, 30–31
  with paste analysis, 124, 130
  paste analysis versus, 147
  potassium, 148
  soil contamination in, 103
  zinc, 112
Tobacco, chloride sensitivity, 97
Tomatoes
  boron tolerance, 101
  calcium requirements, 173
  chloride sensitivity, 97
  manganese tolerance, 105
  molybdenum fertilization, 115
  nitrogen excess effects, 30, 33, 161
  nitrogen requirements, 26–27
  nutrient removal rates, 26, 78, 84
  sodium application, 92
  sodium uptake, 92
  zinc deficiency tolerance, 112
Top dressing, phosphorus content, 43
Total exchange capacity (TEC), 5
  calculation, 14–15
  definition, 13
  exaggerated, 5
  noted on soil reports, 12, 13–15

Total exchange capacity soil, calcium-magnesium balance, 39
Trace elements, 99–120
  amounts needed, 148–149
  banded, 169
  in calcareous soils, 167, 168, 169, 170, 171
  chelated, 191
  dry and foliar applications, 137, 140
  as enzyme catalysts, 103
  foliar feeding, 171, 176
  glyphosate chelation, 175
  heavy metals, 118–120
  in HEC soils, 151–256, 152, 153, 154, 155, 156
  impaired solubility, 173
  irrigation water effects on, 208
  in LEC soils, 137, 148–149
  as Modified Growing Media component, 188, 191
  paste analysis, 128, 130, 141, 144, 146
  SLAN balance approach, 134, 137
Turf grass, 37, 39, 84, 140, 141

U
Urban farming, 218
Urea, 24, 31
Urea-based fertilizers, 28
  side-dressed, 29
Urea-based proteins, 28

V
Vermiculite, 13
Vineyards, soil copper levels, 110
Vitamin A deficiency, 37
Vitamin B, 100
Vitamin B12, 115
Vitamin C, 107

W
Water analysis, 18, 205–214
Water availability, 179–180
  field capacity, 179, 180
  by soil type, 179–182
  wilting point and, 179, 180
Water budget, of plants, 70
Water extraction analysis. *See* Paste analysis
Water quality, 43–45
Water Quality for Agriculture (F.A.O.), 205
Water Quality for Irrigation (Stromberg), 205
Water utilization, 107, 175, 180
Weather
  See also Rainfall
  foliar feeding and, 173–174, 175

nutrient availability effects, 179–182
Weed control, 47, 218
Well water, 191, 209
Wet weather, 181–182
  See also Rainfall
Wheat
  boron tolerance, 101
  chloride sensitivity, 97
  copper sensitivity, 110
  magnesium deficiency, 70
  moldboard plowing, 45
  nutrient removal rate, 26, 78
  silica fertilization, 117
  sodium application, 92
  zinc deficiency tolerance, 112
Wilting point, 179, 180
Wind, 175
Winter wheat, 141–142
Wollastonite, 118
Wood ash, 54, 71
Worm castings, 187–188

Z

Zinc, 137
  allowable cumulative loading rates, 119
  application methods, 144–145
  interaction with phosphorus, 173
  for LEC soils, 137
  mobility in plants, 110, 111
  mobility in soil, 110, 111
  molybdenum uptake effects, 114
  nickel uptake effects, 120
  paste analysis, 128
  phosphorus uptake effects, 142
  soil test, 5
  sources, 113
  tie-up, 173
Zinc carbonate, 113
Zinc deficiency, 107, 110, 111–112
  in LEC soils, 141, 144–145, 147
  paste test, 144, 147
Zinc lignosulfonates, 113
Zinc oxide, 113
Zinc phosphates, 111
Zinc sulfate, 37, 113, 144–145, 151, 158
  foliar feeding, 176
Zinc toxicity, 111

Acres U.S.A. — books are just the beginning!

Farmers and gardeners around the world are learning to grow bountiful crops profitably — without risking their own health and destroying the fertility of the soil. **Acres U.S.A.** can show you how. If you want to be on the cutting edge of organic and sustainable growing technologies, techniques, markets, news, analysis and trends, look to **Acres U.S.A.** For 40 years, we've been the independent voice for eco-agriculture. Each monthly issue is packed with practical, hands-on information you can put to work on your farm, bringing solutions to your most pressing problems.
Get the advice consultants charge thousands for . . .

- Fertility management
- Non-chemical weed & insect control
- Specialty crops & marketing
- Grazing, composting & natural veterinary care
- Soil's link to human & animal health

For a free sample copy or to subscribe, visit us online at
www.acresusa.com
or call toll-free in the U.S. and Canada

1-800-355-5313

Outside U.S. & Canada call 970-392-4464
info@acresusa.com

Learn more from Eco-Ag U Online
Learn.AcresUSA.com

Tap into the knowledge of eco-agriculture's top farmers, ranchers, scientists and consultants from your own home. All classes are designed by expert agronomists and professional farmers, and designed by the educational experts at Acres U.S.A., the Voice of Eco-Agriculture.

STUDENT TESTIMONIAL:

"Eco-Ag U Online from Acres U.S.A. offers an unparalleled learning system for new or experienced farmers. The teachers are world-class communicators and leading thinkers in their fields. The platform creates a very powerful learning management system, easy to access and very interactive, with clear course goals and how to achieve them. It leaves you with a definite understanding of the basic principles. Plus, the courses came with additional learning opportunities from a very large database of free articles from Acres U.S.A. With a self-review test for each lesson, you can assess your comprehension as you go. And as soon as I registered, I got outstanding support from Acres U.S.A. and answers to any questions I had as I took the courses. If you are just discovering the environment degradation from the use of chemicals in conventional farming or a farmer wanting to transition to regenerative practices, the premium package is guaranteed to help you put together your first ecological farm plan. Eco-Ag U Online is your gateway to understanding the principles of eco-agriculture in theory and in practice."

Fabien Peulvast
United Kingdom
Premium Package Student

COURSES CURRENTLY AVAILABLE INCLUDE:

A Growers Guide to Regenerative Hemp

Instructor Doug Fine

Doug Fine is developing a high-end market for regenerative hemp. Learn from the author of multiple books and the operator of multiple hemp farms what it takes to succeed today. From soil management lessons to cultivation techniques, this class will give you the necessary tools to start your small-scale hemp operation and grow it into an industry leader.

    Customer Full Price: $199

    Scheduled launch date: Live!

Practical Permaculture and Agroforestry for Farmers

Instructor Mark Shepard

Learn from world-renowned farmer and best-selling author Mark Shepard how to build your very own commercial permaculture operation. Not just a theory class, you'll hear real-world advice and tips on these innovative farming systems. Whether you "go all the way," as Mark is doing, or simply introduce some new crops and diversity into your traditional farm or small acreage, this workshop is certain to provide the roadmap you'll need.

    Customer Full Price: $599

    Launch Date: Live!

Biological Farming Business: Develop a High-Yield Crop System

Instructor Gary Zimmer

Gary Zimmer, known as the grandfather of biological farming, has created a detailed class to help you build your own biological farming system. From soil management to soil testing to fertilizer recommendations, Gary demonstrates and gives guides students through of how to build and manage a balanced soil system, and how to build a profitable operation.

    Customer Full Price: $249

    Scheduled launch date: Live!

Proven Lessons for Success in the Business of Farming

Instructor Paul Dorrance

Learn from a real farmer and rancher about best practices in setting up your business. From bookkeeping to marketing, making sure you are tracking what you measure with your books is as important to measuring your fields. Includes real-world examples to help you get started.

    Customer Full Price: $149

    Scheduled launch date: Live!